Radio Free Earth

The Complete Beginner's Guide to Survival Communications

Radio Free Earth

The Complete Beginner's Guide
to Survival Communications

Marshall Masters
Duane W. Brayton

Your Own World Books
Nevada, USA

RadioFreeEarth.org
YowBooks.com

COPYRIGHT

Radio Free Earth: The Complete Beginner's Guide to Survival Communications

Your Own World Books
An Imprint of Knowledge Mountain Church of Perpetual Genesis, NV, USA
Author: Marshall Masters

Trade Paperback
Third Edition — July 2021
ISBN: 978-1-59772-195-0
www.radiofreeearth.org
www.yowbooks.com

Related Titles

Win-Win Survival Handbook: All-Hazards Safety and Future Space Colonization

Everything we are taught about surviving the "end-of-life-as-we-know-it" is wrong, according to Win-Win Survival Handbook author Marshall Masters, and here is why:

The conventional Plan A is about building to fail and playing the odds. When the odds do not pan out, Plan B is about huddling in a box in the ground, eating dead food, and wondering what comes next.

With a Win-Win, you build for continuity of life so that you are going over speed bumps when others are hitting walls. Your Plan B is the noble and life-affirming mission to prepare your progeny to colonize distant worlds.

This book guides you through the development process with detailed instructions for designing, building, and shielding communities for self-sufficiency, survival, colonization, and profit.

In a country blessed with Win-Wins, there will always be hope for the future – no matter what comes our way.

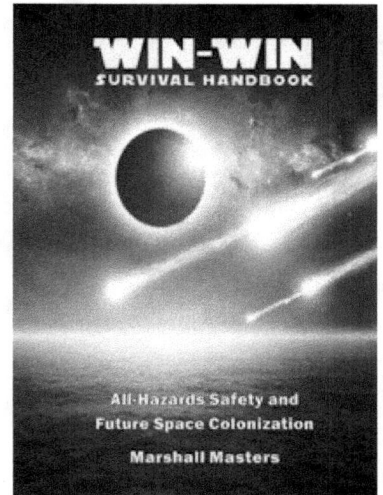

Surviving the Planet X Tribulation: There is Strength in Numbers

Why examine the need to survive a Planet X Tribulation? In a word, history, and here is why.

Astronomer Percival Lowell, founder of the Lowell observatory in Flagstaff, AZ, coined the term "Planet X" around 1905. It describes a sizeable object in space that European astronomers called "Neptune's Perturber." They could not observe it but knew it was there due to perturbations observed in the orbit of Neptune.

Astronomer Clyde Tombaugh of the Lowell Observatory discovered Planet Pluto in 1930, and it was assumed to be "Neptune's Perturber." Decades later, astronomers discovered that Pluto is only 60% the size of Earth's moon. Hence, it lacks the mass to be Lowell's, long-sought Planet X, and the search for continues.

With this in mind, imagine that you live to see Planet X with your eyes. What would you do, and with whom, and more importantly, how will you work together to survive?

The answers to these and many other questions are in this book. Written in an entertaining style, it will gently and positively immerse you in the topic.

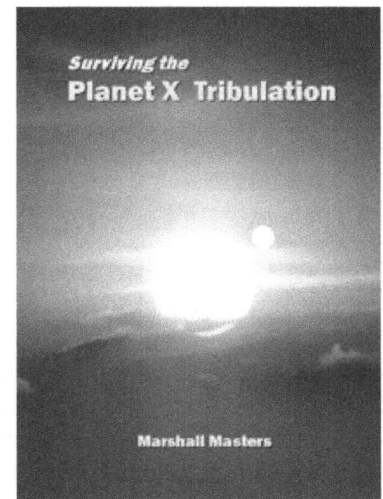

Dedication

Nikola Tesla first invented the idea of the radio in 1892, and the communication technology discussed in this book would have been impossible without his creative genius. It is quite likely that without him, we'd all still be waiting for Western Union telegrams.

In 1901, it was Guglielmo Marconi who used Tesla's work to transmit the first wireless Morse code message across the Atlantic. As a result, the two men found themselves in a protracted legal battle. Six months after Tesla's death in 1943 the US Supreme Court ruled that all of Marconi's radio patents were invalid, and they awarded the patents for radio to Tesla because of Tesla's "Prior Art."

How was it that Tesla's legacy was buried for so many years? Inventor, Thomas Edison was jealous of Tesla's inventive genius and suppressed his legacy causing him to languish in obscurity. The information age changed that, and today, more people cherish Tesla's contributions than ever before; though this is not about a difference of genius. Rather, it is about passion. Tesla's passion was first, last, and always to help humanity; whereas, Edison's first passion was always Edison.

What makes Tesla precious for humanity is that he gave us access to the immense power of the radio frequency spectrum, and today, we work it from one end to the other.

What is the radio frequency spectrum? You'll learn the technical aspects in this book, but from this author's perspective, it represents God's chat room. Through it, humanity can now cross a vital evolutionary threshold into what some would call a "Star Trek future," and it was Tesla who gave us a front row seat. It is for this reason that I dedicate this book to him.

Acknowledgments

This book represents a culmination of over two years of intensive research and authoring. It would not have been possible without the wonderful people who freely contributed their time and talents to this effort.

To Duane Brayton, my heartfelt thanks. A seasoned HAM, with a solid radio and electronic engineering background, his decades of experience and technical expertise were absolutely essential in the creation of this book.

For newbies, the most difficult thing about two-way radio communications is its many complexities. So, our authoring mantra was "We're serving those who see two-way radios as survival tools and those among them who think that a transceiver is a sexual preference." (When I wasn't reminding Duane of this, he was reminding me.)

To Joseph Lewis thanks for his wonderful assistance as the primary editor for this book. Joe's focus was on weeding out the techno babble and smoothing out the text. The result is a more readable book.

To Doug Abramson, my sincere thanks for his assistance with the strategic aspects in the book. Doug was the instructor for my Amateur Radio Emergency Service (ARES) certification course.

Above all, my deepest thanks to all of you who have supported my work through the years.

Table of Contents

1

Introduction

Is a single question dominating your need-to-know priority list, such as, how much is this going to cost me? If so, congratulations, you're off to a good start as you're likely not a HAM.

HAMS, amateur two-way radio operators, are passionate hobbyists and experimenters so the question they are more likely to ask is "how does it work?" Cost comes later. First is the passion and they use the term "the bite," to describe it.

When you get "the bite," it1. means that you've become passionate about the technology, and here is the truth of it. If you want to learn absolutely everything there is to know about two-way radios, you'll need the patience of a Buddhist monk because it will take you at least three lifetimes to do it.

However, if you're not a HAM, the bite is irrelevant because you need information, the kind that helps define what is possible and what is necessary when using two-way radio technology for survival. For this reason, a business-like strategic plan offers the most direct and cost-effective way to achieve this aim, and in this chapter, we'll discuss the following seven different strategic planning goals:

- **Buying Radios for Survival:** Why this is a great time to buy two-way radios for preparedness applications.

- **Community Communication Strategy:** This book employs a bottom-up, three tier strategy for the acquisition and prioritization of reliable two-way radio technologies.

- **Technical Team:** Leaders must focus on human engineering. Engineers focus on radio engineering, and several skill sets are discussed.

- **HAMS and Appendices:** Licensed amateur radio operators (HAMS) will be the front line of your system. Finding them and supporting them are vital.

- **Morse Code Proficiency:** Morse code is essential because it remains the most effective way to transmit messages in adverse conditions and over long distances.

- **Alliance Partners:** Working with local municipalities and other survival groups offers strength through numbers and mutual support for survival and defense needs.

- **Radio Free Earth:** Starting your own local Radio Free Earth (RFE) network with the support of your alliance partners.

The seventh and final goal is to organize a local Radio Free Earth network and the first six will help get you there. The rest is everything in between and with two-way radios the immediate concern for survival community leaders is cost.

Buying Radios for Survival

The assumption here is that you do not have the bite with two-way radios, so the question, "how much is this going to cost me?" has more than one meaning.

- **Human Cost:** The first and most important is to answer what is the human cost. How many graves you're willing to dig because you failed to plan properly and didn't see or hear about trouble coming in time to respond?

- **Wasteful Experiments:** How much are you willing to spend on needless trial-and-error experiments that will drain your finances because you are lazy or shortsighted?

- **Priority:** How much you're willing to spend is likely to be an issue of priority as opposed to resources. Ergo, if two-way radios find their way to the bottom of your priority list, that's how they'll perform when you need them most.

In the final analysis, if you are diligent in creating a strategic plan, no matter how much you can afford to spend, it must be put to good use. That's the goal of this book. To help you do exactly that and timing is important.

Obviously, you will have more available options until folks in the mainstream begin to perceive the signs of an impending global tribulation and so, take action. Once that happens, your range of options will narrow considerably. This is why the sooner you can acquire your two-way radios, the more options you'll have and the less expensive they'll be.

Granted, this can be daunting. For example, you look at advertisements for two different AR-15 sporting rifles. One is priced in the hundreds of dollars and the other in the thousands of dollars. To the untrained eye, there is not much difference, save for the price tag. For those who understand the difference, the two rifles appear as different as night and day.

At short range, a head shot with an AR-15 sporting rifle costing hundreds of dollars and one costing thousands of dollars is a head shot. Bang, you're dead. But what about long range shots? Here is where the difference between hundreds and thousands of dollars comes in with added range, reliability and functionality.

The same applies to two-way radios. You can find yourself looking at two comparable offerings, one priced in the hundreds of dollars and another priced in the thousands of dollars. To the untrained eye, both can look equally impressive at first glance. For those who know what they're shopping for, each two-way radio can be impressive in its own right. Therefore, which do you choose? The one that is right for the survival mission, and this does not include the latest state-of-the-art two-way radios available. This is because the radio industry is in the process of converting from older, traditional analog designs to newer, more expensive and complex digital designs that essentially combine a radio and a computer into one component.

It may seem logical that these new state-of-the-art digital components are ideal for survival applications, but this is not the case. Newer is not necessarily better when it comes to survival. For this reason, this book recommends the best value-proposition which is using the last generations of popular, state-of-the-art analog two-way radios, for the following reasons:

- **Lower Cost:** Most digital two-way radios sold today are relatively new and manufacturers must price their digital radios to recoup their development costs. Conversely, popular analog designs that are a decade old or longer are less expensive in part because their design costs have been fully amortized.

- **Time Tested:** Analog two-way radio models that have been in the market for a decade are proven designs. This means there will be a lot of them available in the used market as well as a number of aftermarket enhancements.

- **Effective Range:** Digital models transmit in narrow frequency ranges. Analog two-way radios transmit in much wider frequency ranges. In times when the range and signal clarity of a two-way radio is reduced by environmental issues, the wider frequency range of an analog model means that it will usually outperform a digital model under the same extreme circumstances.

- **Compatibility:** At present, manufactures are in a tug of war with each other to see who will set the digital standard for the industry. Consequently, while two digital radios by different brands may may look the same, there will likely be compatibility issues until the industry agrees to one or more universal standards. With analog two-way radios, design, regardless of when it was introduced or by whom, will always be compatible on the same licensed frequencies.

- **Infrastructure-Independence:** Since the earliest days of wireless Morse code transmitters, two-way radios have always been somewhat infrastructure-independent. Regrettably, newer digital designs are nearly completely infrastructure-dependent on specialized networks and the Internet.

The first four reasons above are more than enough to justify the use of older analog designs, but infrastructure is the proverbial 800-lb gorilla in the room. This is because digital designs are infrastructure-dependent, and during a global tribulation, many of these complex and expensive features will have nothing to inter-operate with. This fact runs contrary to a noble history.

When disasters strike and infrastructure fails, this is when commercial media and communications systems go dark; during a global tribulation, they'll stay dark. This is when HAMs become, as former FEMA Administrator Craig Fugate stated, "our last line of defense."

Such was the case in 2005 with Hurricane Katrina, when HAMs across the country packed their gear, fueled their trucks and headed South into that devastated region. Thanks to their efforts, aid was directed to where it was needed, families were comforted and more. This was not a one-time event. It is part of an ongoing legacy of public service by the members of The American Radio Relay League (ARRL). Founded in 1914, their motto has always been, "when all else fails... amateur radio."

So what does this all boil down to when buying two-way radios for survival? It is the Goldilocks principle, from the popular short story, Goldilocks and the Three Bears, where a little girl named Goldilocks tastes three bowls of porridge. One is too hot. One is too cold and one is just right.

In terms of two-way radios, the hot ones are expensive and digital, whereas the cold ones are cheap and unreliable. What's just right? Popular and reliable analog designs that have been on the market for at least five years and which are value-priced. When you finish reading this book, you'll know how to spot all three and what to do with the ones that are just right.

With this in mind, you need to start building a vocabulary of terms, beginning with commonly- used, function-specific terms used in lieu of "radio."

- **Receiver (aka radio):** A radio that can only receive and play transmissions.
- **Transmitter:** A radio that can only send transmissions. In the early days of amateur radio, HAMs typically used two separate radios: a receiver and a transmitter.
- **Transceiver (aka two-way radio):** A dual function radio that works as a combined receiver and transmitter.

From this point forward, these industry standard terms and others we'll discuss along the way, will be necessary when shopping for communications equipment. Also, a basic vocabulary is necessary for creating a community communications strategy based on your needs and aims.

Community Communication Strategy

No two survival communities will have the same exact communication requirements, so there are no "one size fits all" solutions. For this reason, the community communication strategy in this book is offered as a reference to help you design and implement your own strategy.

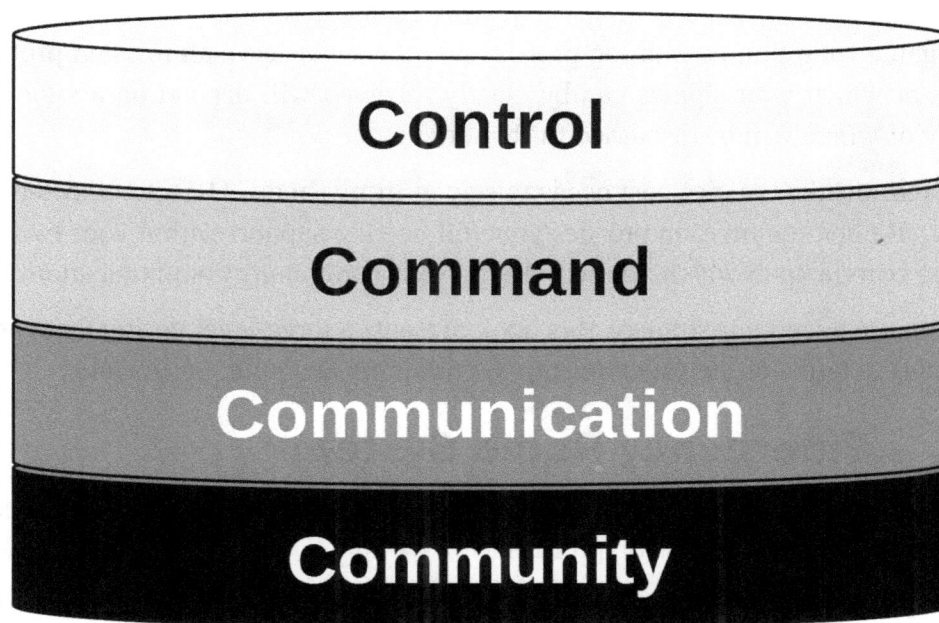

Whatever your eventual strategy becomes, you will enhance its chances of success by building it to serve the "four C's:" (C)ommunity, (C)ommunication, (C)ommand and (C)ontrol. This requires a strategy that will make sense with the two perimeters that will define your operating and security missions, your community's inner and outer perimeters.

For the purpose of this reference example, the inside perimeter will be a tree line that is one hundred and twenty five yards away from the community structures. Between the structures and the tree line, the field is leveled and bare to offer a clear killing field for attacks that penetrate the inner perimeter.

Needless to say, if an attack penetrates your inner perimeter, members of your community will die and in some cases, very badly. Therefore, dealing with attackers before that, requires an outer perimeter.

With an outer perimeter you are defining a killing zone which will likely be beyond your line of sight such as a stream bed at the base of the reverse slope of a mountain bordering your community. It could likewise be even further than that.

From a two-way radio perspective, the area inside your inner perimeter is primarily going to require low-power, line of sight communications. Beyond that is the outer perimeter and beyond that, no man's land. In these situations, there will be line-of-sight communication missions and there will field operations where your teams are operating well beyond your outer perimeter.

The mission profile for operations outside of your outer perimeter also will need to extend their messages and instructions over one or more mountain ridgelines or summits. This will require different types of long distance two-way radios as opposed to line-of-sight only.

Another factor to consider is overlap. Two-way radios do not have fixed operating ranges. Some are suitable for a limited mission profile and others can serve all mission profiles. Ergo, the distances at which your signals can be clearly received will depend on a wide number of factors, many of which will be discussed further on.

For these reasons the issue of cost needs to stay in firmly in the TBD (to be determined) column for now. Rather, the mission profiles you will need to support within your two perimeters need to be the criteria upon which you will design your community communications strategy.

To help you organize your strategy, this book presents a three-level design that encompasses mission profiles in terms of their relationship to your inner and outer perimeters.

Level 1 – Emergency Radio System

Level 1 will be first part of your communication system to be designed and implemented. It will serve the following three mission profiles:

- **Community Operations:** The principal mission profile for level 1 will be day-to-day, routine, operational line-of-sight voice communications, inside the boundary to the outer perimeter of your community.

- **Regional Emergencies:** In situations where the community can sustain itself during a local catastrophe, you'll need to support line-of-sight and long distance (DX) voice communications beyond the outer perimeter of your community.

- **Long Distance Communication (DX):** Small, inexpensive, two-way radios that only offer CW (Continuous Wave Morse code) communications will serve two roles. First, they will help community members develop their Morse code skills with general long distance traffic.

  The second role will be to address the worst possible scenario for your community where your structures are destroyed, and your members are scattered in small groups and at far distances. In this mission role, the same inexpensive CW two-way radios used for Morse code training and proficiency will be used by community members to coordinate, collaborate and share information at distances of up to several hundred miles.

The two-way radios you choose for this first level of your community communications strategy represent the foundation of your system, and they will be used in levels 2 and 3 as well.

Level 2 – Mobile Field Operations

Level 2 mission teams will principally operate beyond your community's outer perimeter. These teams will include foraging, rescue, defense, and surveillance. They will require multiple types of rugged, high-power, line-of-sight and long distance two-way voice radios and repeaters.

- **Security and Reconnaissance:** The principal mission profile for level 2 will be long distance (DX) command and control voice communications in support of sniper and reconnaissance teams operating beyond your outer perimeter with DX two-way radios and/or repeaters.

  For sniper teams and observation teams operating between your inner and outer perimeters, the component mix will lean more towards less complex, line-of-sight voice communications.

- **Scavenging and Rescue:** The principal difference between this and the security and reconnaissance profile will be the operating environment. With security and reconnaissance, communications will take place over wide areas of relatively unobstructed terrain, and this requirement also applies to scavenging and rescue teams, with a caveat.

  Scavenging and rescue missions will also require the ability to communicate through man-made structures as well as across large areas of unobstructed terrain. Expect these difference to require a different mix of components for these very different mission profiles.

- **Local Intelligence and News Gathering:** The mission profile for your intelligence and news gathering teams shares a similar requirement. These teams must be able to work with as many different amateur and consumer radio frequencies and services as possible and with the greatest degree of flexibility. Public presence is where these two missions differ.

  Your intelligence teams will operate locally in a covert manner or with a minimum of public presence. Their principal activity will be to observe ground movements within an assigned sector and to gather signals intelligence from line-of-sight consumer and amateur radio services.

  On the other hand, your news gathering teams will operate with a high public presence in support of your Radio Free Earth broadcast operations.

When budgeting for acquisition and stockpiling, level 2 will demand the largest share of your community communications strategy budget, even though more physical units will be needed for level 1. This is partly because of the degree of integration required between your level 2 and level 3 mission profile requirements.

Level 3 – Operations and RFE Broadcasts

Level 3 will require the most expensive and flexible components, all of which must be proven, popular and infrastructure-independent. One advantage with level 3 is that it uses the kinds of components that are found everywhere in the radio rooms of licensed HAM operators. How many HAMs are there in the USA? While not all HAMs are members of the American Radio Relay League (ARRL), at present there are over 160,000 active ARRL members and a significant number of them hold engineering degrees.

This is a critical fact. While you'll work primarily with skilled engineers and technicians to build your three-level communication system, the men and women who will man your level 3 base stations will likely have a call sign and current membership with ARRL. Here are the most important missions they'll be performing:

- **Command and Control:** This is the primary mission role for level 3. All command and control radio traffic is deemed to be a high-priority stakes role which takes precedence over all other level 3 mission roles and profiles.

  All level 1 community members and level 2 field operations teams are monitored and commanded by the community leadership at level 3. This includes all scheduled and unscheduled events and activities.

- **Voice/Data/CW Traffic:** Two-way communications include several different modes such as voice (using a microphone), data (computer generated messages) CW (Continuous Wave) Morse code. Unlike command and control, this mission is for normal priority traffic and allows operators to experiment with different communication modes and methods.

  With data traffic, operators can transmit data files, image files, faxes and video using amateur radio frequencies. With CW traffic, operators proficient in Morse code will make local and long distance contacts at ranges of up to 2,000 miles or more.

- **DX Listening Stations:** One of the level 2 mission roles discussed above is Local Intelligence and News Gathering. Both of these mission roles are primarily intended for monitoring local amateur and consumer radio frequencies.

  With level 3 DX (long distance) listening stations, all forms of two-way radio and broadcast radio traffic are passively monitored for intelligence. The goal of these DX listening stations is to identify active HAM and broadcast stations, so that the leadership can evaluate possible contact opportunities.

- **Security Alliances:** By definition, a global tribulation is a multi-year event lasting up to a decade or perhaps longer. During this time, federal and state governments will come under heavy pressure from dwindling resources. Hence their presence will diminish significantly over time, if not altogether. In this case, we'll see the re-emergence of small city states, and here is where you'll need to build alliances with the elected officials of local municipalities and public safety services.

Here, you need to have a set rule. You only deal with *elected* officials which means you'll want to work with elected Sheriffs – not appointed city police chiefs. Also, here is where your security leadership must be the ones to manage alliance traffic on a daily basis.

- **Radio Free Earth (RFE):** The interesting thing about this mission is that while it will be the last to be implemented, over time, Radio Free Earth broadcasting will become the most important knowledge resource for the long-term viability of your mission, your community, the alliance communities, and alliance municipalities.

  Initially, you'll broadcast using a consumer service such as Citizens Band but with considerably more power than is presently authorized by the FCC. From there, you'll draw upon your alliances to locate an abandoned commercial radio station for broadcasts on a commercial AM or FM frequency.

- **Support and Training:** The last thing you want to do is to begin learning how to operate a radio in a moment when you need it the most. Your chances of success will be considerably more effective with consumer class two-way radios, but with amateur radios, you'll hit walls faster than you could begin to imagine.

  This is why you'll need to train community members on how to use the two-way radios and how to send and receive Morse code messages via CW, because this is the most reliable and powerful way to establish two-way conversations with HAM operators locally, regionally and internationally.

You may wonder why it is beneficial to start a local Radio Free Earth network with active news-gathering and regional broadcasting. The answer is global tribulation choice. You can choose to be a part of the solution, a part of the problem or a victim.

If you're a handful of preppers lurking about in the woods with sniper rifles costing thousands of dollars and saying clever things like, "If I see you in the cross-hairs of my scope, you're mine," the local sheriff is going to see you as a part of the problem, and unless you're dealt with early on, there will be more victims to take care of – not less.

On the other hand, if you walk into a sheriff's office and tell them that you're starting a local network for a public service role and that you want to forge an alliance with them, they'll be interested. If you walk in with local HAMs who've volunteered to work with you, an honest sheriff looking for honest solutions will be doubly-interested.

You now have a high-level overview of the three-level design. Future sections and chapters in this book will flesh out the rest of the technical information. Your immediate concern will be human engineering, not radio engineering. For that, you'll need to build a technical team.

Technical Team

HAM operators often perform a number of engineering tasks to build their own systems and they can be quite resourceful. However, as a leader you need to take a broader system engineering view, and here is where you want the knowledge and work experience of skilled engineers.

There are seven skill sets you need to look for, so think of them as seven different hats hanging from a hat rack. At the outset, you may have just one or two people to wear all seven hats, but over time, additional recruitment opportunities will occur. When they do, you'll know what you're looking for so that you can seize these opportunities with confidence. After that, trust the wisdom of Sun Tzu, the author of *The Art of War*. "Opportunities multiply as they are seized."

Keeping this axiom in mind, the first technical team opportunity you must seek and seize, will be the recruitment of a team leader; someone, ideally, who is a warm, patient, and genuine problem-solver with relationship-oriented communication skills.

Production Engineer / Team Leader

As a rule, you're not going to see a lot of engineers attending a Toastmasters Club public speaking course, because most only have enough people skills to order a pizza. Beyond that, they prefer to work in a world of their own and not to be disturbed. However, production engineers are unique, because they tend to be people-oriented problem solvers.

A production engineer's responsibilities cover the whole gambit of the manufacturing process from the back office to the front office. As buffers between the engineers and the eventual users of the product, they translate customer desires into specifications engineers can work with and oversee the production, delivery and implementation of the product.

In a commercial, for-profit world they are seen as necessary facilitators. However, in terms of implementing an emergency radio system and more sophisticated news gathering and operations, your production engineer will be the living incarnation of the Marine mantra, "Improvise, Adapt and Overcome."

This is why the person you need to be the leader of your engineering team should be an experienced production engineer or someone who fits the same mold.

Systems Engineer

Most engineers work on a specific component of a system. What they see is a micro view of what is necessary to make that component serve its role within a greater system. Conversely, a systems engineer has a macro view of an entire system that encompasses all of its components.

Technically speaking, a systems engineer is a big picture engineer who uses an interdisciplinary approach to both engineering and engineering management. Another way to explain it is to use an old axiom that is less technical.

How do you climb a hill? One step at a time. In other words, a systems engineer can visualize the whole hill as a complete system and then how to climb it one step at a time and in which order.

By having an engineer with this ability on your team, regardless of whatever resources are available to you, your communication systems will evolve in a predictable and reliable way.

Radio Engineer

Radio engineers are also referred to as radio frequency engineers. They are electrical engineers who specialize in working with devices that emit and receive radio waves, and work with a broad range of devices to include radios, wireless networks, and cell phones.

During the global tribulation, the ability of your network to send and receive radio transmissions will be plagued with a wide number of environmental issues, to include airborne particulates, solar weather, radiation and so forth. Consequently, things that work reliably today may not work reliably in this new environment.

Meeting these future challenges will require the ability to think outside the box, and here is where a radio engineer can bring added-value by providing input based on this specialty subject knowledge to your engineering team.

Software Engineer

The next time you use a smart phone app, run a database program on your company's mainframe computer or use a personal computer, laptop or tablet at home or the office, it'll work because a software engineer made it work.

A software engineer designs, develops and tests the software used by any programmable device. These devices can be computers used by your HAM operators to control their radios, but they can also be the programmable hydroponics controllers you will use to help you feed your survival community.

Hardware Engineer

The counterpart to the software engineer is hardware engineers. They design, build and work with components such as processors, circuit boards, memory devices and networks.

For example, a hardware engineer can build a programmable hydroponics controller that continuously monitors the system so that disruptions such as leaks, blockages and pH imbalances can be detected before they have a chance to cause a serious system failure.

In this example, it will be the systems engineer who envisions a complete integrated control system for your hydroponics farm and then directs both the hardware and software engineers to develop the components of a reliable system.

To really appreciate how much you need a hardware engineer, ask a ham radio operator to open up one of his modern radios so that you can look inside. When you do, there will be no clunky parts and vacuum tubes. Rather, you will see that the guts of a modern ham radio look very similar to the guts of your own personal computer.

Antenna Guru

The tribulation is going to be especially brutal on antennas. Hard blowing dust, debris impacts, hyper velocity winds, super hail and so forth will comprise some of the worst possible environments.

This is why the best thing that could possibly happen to you is to find some talented teenager who loves nothing more than to make and tune radio antennas. And do not be surprised to learn that your candidate doesn't hold an FCC license and call sign. That's OK. What matters is passion.

All your engineers need to do is to tell your antenna guru what they want and then get out of the way. That is when they'll watch in amazement as bits from one salvage and parts from another are gleefully cobbled together to make an antenna to the exact specification. And likewise, sometimes your engineers will get more than they could have dreamed for.

Systems Analyst / Senior Technical Writer

Technical writing, as with other technical fields, has its own rankings and specialties. Most technical writers employed today produce what is known as user documentation. They write the instruction books that come with your computer and software.

WHEN paper, toner, ink and the power to run office printers fall into short supply, the need for documentation will be even greater. This is when USB thumb drives, smartphone memory cards and other storage devices will become your new bookshelves. Not just for technical documentation, they will contain a wide range of fiction and nonfiction literature as well.

However, when building your communications systems, you will need the kind of writer who can analyze and understand systems and then explain them to people with low technical skills. For this, you want a specialized writer known as a systems analyst / senior technical writer with no less than five years of work experience.

In the field of technical writing, these are the ones with the highest ranking who will take global ownership of a documentation effort and see it through from analysis to end-user training.

Building Your Engineering Team

Up to now we've looked at seven different skill sets, and this is just a partial list. You will also need repair technicians, field installers and so forth.

As you organize your community, you will likely begin with just one or two people who will have to wear many hats, which is the point of this part of the book. If you can identify the kinds of hats people will need to wear in order to make your communication strategy work, you'll eventually come across people willing to wear those hats.

When you do, you'll understand where they fit into the scheme of things and how they can help you be successful with your communication strategy. Most importantly, you'll be able to acquire the equipment you'll need while minimizing expensive learning curve mistakes.

At the outset, you and your engineering team will need to identify the kinds of radio transceivers and supporting gear your survival community will need to serve a wide range of needs.

To achieve this, you will need to lead radio operators and engineers by employing human engineering, your most important leadership skill. While your technologists are connecting cables and gear, you'll be connecting with them with the Three I's: invention, insight, and initiative. For this reason, you need team players who are willing to take ownership of a vision.

As you evaluate potential technical team candidates, their credentials, experience and skill should be held in balance with personal strengths. They should be naturally inventive people with ability to study a problem until they form an insight. And above all else, they should be able to take initiative. These are people who proactively assess things independently and then act before others do.

What you want to avoid are people who define themselves by relying on negative contractions. Using can't, wouldn't, shouldn't, couldn't, isn't and other such words are reasons to fail.

Overall, we've established your strategy mission roles and identified the skill sets you'll need to build a successful technical team. Yet, everything you've done to this point has been to flesh out the back ranks of your community communications strategy. Now, we need to focus on who will serve in the front rank positions. Namely, we mean the HAMs and their apprentices.

HAMS and Apprentices

When you turn on a receiver or transceiver to listen for traffic, you'll hear others on the air calling for others to dialog with. For newbies, it is exciting to hear someone over the air, and the urge to answer and start a conversation is strong. What a HAM will tell you is to ignore your urge because there will be operators on the air you'll want to talk with and others, you must never talk with. Here is where a HAM's many years of on-air experience becomes invaluable.

For this reason, HAMs will be the front line of your operation; most all of today's HAMs are seniors. Attend a local radio club meeting and you'll see it with your own eyes. If there are one hundred HAMs in the room, you'll be hard pressed to find participants with hair on their heads. The rest (like Silver Back Gorillas) are a dying breed, and when they pass, the term a HAM would use is they become a "dead key." This is a term that dates back to the early Morse code days of radio. So where are the young people (presumably with hair on their heads)?

Before the advent of cell phones, there was always a steady generational stream of new entrants into the amateur radio world. But smartphones changed that because the ability to have a computer in one's hand changed everything. Now, a smartphone can connect you to the world in a myriad of ways. Hence, the future HAM interest of post baby boomer generations is feeble at best.

As to the women, most are involved in amateur radio because of their husbands. Nonetheless, during the coming tribulation, these older men and women who now dominate the amateur HAM radio population will become your global tribulation human treasures, especially when you begin building your local Radio Free Earth network.

As the leader of a survival community, you may hear some members of your community complain about the care and feeding of these old HAMs who, in some cases, will not be strong enough to pull weeds. If so, be patient with those complaints but do not yield to them because during the tribulation, survival will be less about the having of things and more about the knowing of things.

Will most HAMs participate when invited? Absolutely, and they will be eager to do so, because they'll want to help and do something purposeful and useful with knowledge they gained over the years. Especially when they can pass that information and wisdom of experience down to a younger generation, and here is where you must begin planning for the future from day one.

Once you've developed a viable community communications strategy with the assistance of your technical team, the next task is not only to find local HAMs to form your front line. You must also pair them up with young teens who will become their radio apprentices and 24/7 companions. This will not be a 9-5 job for these teens. This is about immersion in the way that only a traditional apprentice program can provide, such as those of the industrial age.

You must choose these young apprentices wisely for there will be those who will see an apprenticeship as a gold brick road out of more mundane and dirty survival roles. The ideal apprentice candidates must have:

- The support of their families;
- "The bite," and not just a casual interest;
- A HAM willing to take them under their wing.

Be careful, for some candidates will only see an opportunity to get out of less glamorous jobs; they could make any promise or claim they'd assume you want to hear. But then again, if you're good at human engineering as leader, you'll be able to quickly flesh out your front rank.

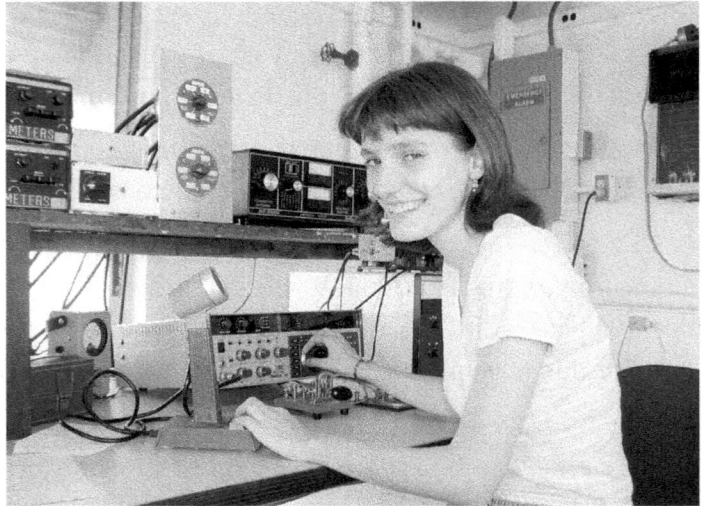

Granted, all of this will likely be a messy process at first, and no doubt there will be days when progress is measured as two steps forward and one step back. In frustrating times such as these, you must always be the one to rise up like a grizzled old Marine Gunnery Sergeant and growl, "OK people. Improvise, adapt, overcome."

Keep sight of your vision of success forever before you, and that is exactly what those you lead shall do. They will improvise, adapt and overcome.

In the next part of the book, Radio Basics, you will greatly expand your radio vocabulary with three chapters: "Modes and Modulations", "Frequency Bandplan", and "Radio Shopping Basics."

When you finish the Radio Basics section of the book, you will possess the vocabulary and technical insights you'll need in order to communicate your ideas and goals to your technical team, HAMs and their apprentices and to manage them on an ongoing basis.

Part 1 – Radio Basics

2

Modes and Modulations

In the previous chapter we discussed a layered community communications strategy to include an engineering team and the need for an emergency radio system. Selecting the right types of radios to use with your community communications strategy depends on a basic lexicon of terms.

In this chapter, we'll begin looking at the basic radio terms and concepts necessary to facilitate the strategy, beginning with modes and modulations. Modes and modulations are the things a radio can do on a given radio frequency. Then in the next chapter, we'll see how radio frequencies are organized.

After you've read this and the next two chapters in this part of the book, you'll be ready to go shopping for radios. We'll do that by evaluating various transceivers offered by manufacturers and how they can be used for a specific survival requirements. However, the biggest initial challenge to newbies is confusion, so this is where we start.

Connecting the Dots

Let's not nibble around the edges. When it comes to amateur radio, there are a lot of things that do not make sense at first glance and which force newbies to run a gauntlet of confusing numbers, letters, codes, abbreviations and acronyms. Consequently, it can take newbies several months if not a year or more to connect the dots on their own.

With that in mind, a venerated and respected radio term you need to know is "Elmer."

Elmers are the seasoned amateur HAM radio operators who take newbies under their wing and gently immerse them into the world of amateur radio. In the process, they spare them a needless waste of time, effort and money. Consequently, those blessed with an Elmer can connect the dots in much less time.

Why is this? Simple. Pose the exact same question to ten different experienced HAM radio operators and you'll get twenty-eight answers. And they'll all be good for the most part.

As the author, I'll to do my best to Elmer you through this overwhelming and expensive gauntlet of confusing numbers, letters, codes, abbreviations and acronyms. It is one that has less to do with technology and is more about the politics of who gets what and where you can use it.

Why are the politics necessary?

Radio Frequency Regulation

The radio spectrum is the radio frequency (RF) portion of the electromagnetic spectrum, and different parts of it are suited to different purposes.

Were it not for governmental regulation, anarchy would prevail over the airwaves. This is why government agencies, the Federal Communications Commission (FCC) and the National Telecommunications and Information Administration (NTIA) are needed to regulate access for the benefit of all.

Access to these frequencies is assigned with an official band plan. This way, the radio spectrum can serve a wide range of governmental, commercial, military, citizen and amateur needs in an orderly and predictable way. Regulatory agencies do this in different ways:

- **Radio Manufacturers:** Radios are designed to work in compliance with federal regulations. Consequently, what you can and cannot do with that radio is determined by government guidelines, which vary from one country to the other.

- **Licensed Frequencies:** Frequencies available to amateur HAM operators require a license for transmitting. However, a license is not required to purchase amateur two-way radios. With consumer two-way radios, licensing can be required as is the case with General Mobile Radio Service (GMRS) radios.

- **Unlicensed Frequencies:** Radios designed for unlicensed frequencies can be freely purchased, and used straight out of the box to both transmit (TX) and receive (RX) without the need to obtain a license. The unlicensed frequencies include Family Radio Service (FRS) and Citizens Band (CB).

Whether you're working with licensed or unlicensed frequencies, there are universal rules and while the FCC enforces different sets of rules depending on frequency and license, there are three things you always need to keep in mind.

- **Receiving Transmissions:** All transmissions on regulated frequencies are transmitted in the open. No matter the radio or the frequency, you do not need a license to listen on any frequency. This is why you will often see radios capable of receiving transmissions on many more frequencies than they are able to transmit.

- **Stay Within Designated Power Limits:** The FCC sets power limits for frequencies, which can also vary depending on licensing restrictions. If you exceed a power limit, you will create reception problems for other stations and neighbors. The FCC will track you down, and they will not be smiling at you.

- **Foul Language:** Regardless whether you are transmitting on a licensed or unlicensed frequency, foul language is never allowed. Some may think that the Citizens Band (CB) is an exception, but they are wrong. FCC Part 95 clearly states: "§ 95.413 (CB Rule 13) What communications are prohibited? To transmit obscene, indecent or profane words, language or meaning."

- **No Secretly Coded Messages or Encryption:** Except for Military and other authorized Government operations, the FCC expressly forbids sending secretly coded or encrypted messages. However, during a global tribulation and the absence of government enforcement, encrypted and encoded messages will become routine.

Still the same, the FCC rules that most licensed HAM operators faithfully obey will endure as a legacy of respect and cooperation among HAMs even with the demise of the FCC. Therefore, embracing this legacy will enhance the reputation of your local Radio Free Earth network.

Now let's imagine that you're shopping for radios online or visiting a radio store. Either way, you are going to be presented with a large range of offerings but the first place to start is to gain an understanding of modes and modulations.

A mode is a way or manner in which something works and with amateur radios, they are restricted or allowed by the FCC on a frequency by frequency basis.

Morse Code (CW)

Morse code was the only way to communicate for the first decade or so of wireless radio. It is based on landline telegraphy of the 19th century to include codes, abbreviations and procedures. When all other modes and modulations fail due to adverse environmental issues, Morse code modes will be your option of last resort.

- **Spark Gap:** Spark gap transmitters offered the first wireless mode for sending Morse code and were the standard technology for the first three decades of radio (1887 to 1916).

Unlike modern CW transceivers which transmit on a single frequency, earlier Spark Gap transmitters radiated "dirty" wide band signals over a wide range of frequencies and thus interfered with transmissions on nearby frequencies.

While the FCC presently prohibits the use of spark gap transmitters, during the global tribulation, a spark gap transmitter could mean the difference between rescue or death. For this reason, at least one member of each survival community family should know how to build a spark gap transmitter in the field after the collapse of our government.

- **Continuous Wave (CW):** The most popular mode for sending and receiving Morse code today is CW. It offers the most efficient use of radio spectrum for sending Morse code messages over great distances.

  No other mode or modulation is more efficient for sending messages via radio than Morse Code CW. In fact, if all other modes fail and you cannot get a message through with Morse code, it's time to build a signal fire.

No matter who you are or what kinds of radios and frequencies you have, the one mode that will prove to be a universal survival constant is CW. It will be the heartbeat of your emergency radio system.

But not only will you be transmitting Morse code, you will also be transmitting voice and perhaps data as well, and here is where the evolution of CW technology sets the stage for a robust number of options.

CW Evolution

When spark gap transmitters were used, they radiated "dirty" signals that propagated across a wide range of frequencies. When just a few radio operators were working Morse code, this was tolerable. However, as more operators began sending more Morse code messages over the air, the need for a more efficient use of the radio spectrum became necessary.

With the advent of practical vacuum tube technologies, more efficient transmitters became possible and they eliminated the problem of sending "dirty" signals. The reason is that they first introduced the highly efficient continuous wave (CW) mode which is still in use today.

With CW mode, a continuous wave is transmitted on a specific frequency. The dits and dahs (dots and dashes) of Morse code are then transmitted as interruptions in the continuous wave. Assuming the sending and receiving transceivers support the CW mode, the dits and dahs that are transmitted as interruptions are converted into tones the operators can hear.

When compared with spark gap transmitters, CW mode is a significantly better second generation technology for Morse code and in fact, remains so to this day.

However, not only was CW mode a blessing to Morse code operators, it also became the enabling technology for radiotelephony, voice over radio or simply "phone" as HAMs call it.

At this point, you may wonder, what does a carrier wave sound like?

A good way to hear a live carrier wave is to listen to a commercial radio station that comes in with a strong signal. Before and after a song plays, there is often a brief moment of silence.

That is the carrier wave. If the signal is weak or subject to interference, you may hear a slight hiss, but otherwise, one can imagine a carrier wave as the sound of silence.

Given that a carrier wave is silent or near-silent, how are the sounds and voices of modern AM and FM commercial broadcasts carried through the air on a silent carrier wave? In a word, modulation.

Radiotelephony Modulation Modes

Voice broadcasting over radio frequencies was formerly called radiotelephony. Today, it is more commonly called "phone" and constitutes the most popular use of two-way radios.

Whether listening to a broadcast AM or FM station, the sounds you enjoy are carried on the sidebands, and are called the "intelligence." The intelligence is paired by the transmitter with a carrier wave (a.k.a. baseband) through a process called modulation. The modulated signal is then transmitted and the receiving radio uses a process called demodulation to remove the base band so the intelligence can be heard.

The sideband above the carrier wave (baseband) is called the Upper Side Band (USB) and the one below it is called the Lower Side Band (LSB). Ergo, a complete signal that brings music, news and commentary to your AM or FM radio is a composite of a baseband and its associated sidebands.

Amplitude Modulation (AM)

Amplitude Modulation (AM) is the oldest form of phone modulation, and work began on AM modulation in the early 1900s. In the 1920s, AM broadcasting was established. AM modulation is accomplished by varying the amplitude of the carrier wave. A secondary effect of this type of modulation is that both sidebands are created which consumes additional power.

All radio transmissions propagate in two ways, the degree of which is determined by the band. The first is Groundwave propagation which bends the signal over obstacles like hills, and travels beyond the horizon. The second is Skywave propagation. Also referred to as "skip," this form of propagation bounces radio waves off the ionosphere.

Although AM signals can be received hundreds of miles away, this modulation has three weaknesses: It is susceptible to interference, offers limited audio fidelity and uses excessive amounts of transmit power when compared with frequency modulation (FM).

Frequency Modulation (FM)

Invented in 1933, frequency modulation (FM) is primarily used in VHF and UHF radios and is mostly line-of-sight transmission; transmissions are usually limited to a range of 1 to 50 miles depending on equipment and circumstances.

FM is the most favorite mode for local use by amateur HAM operators. When you see HAM radio operators with handheld radios called Handy-Talkies (HT), they will typically be FM transceivers with the short flexible antennas supplied by the manufacturers, called "rubber ducks."

While FM may use more bandwidth than AM, it typically offers superior sound quality and is less susceptible to interference. It also uses transmit power in a more efficient way than AM. Where FM and AM are similar is that both modes use a baseband carrier wave.

There are several differences between the two modes but the visual difference between AM and FM modulation as seen on an oscilloscope or spectrum analyzer is more revealing.

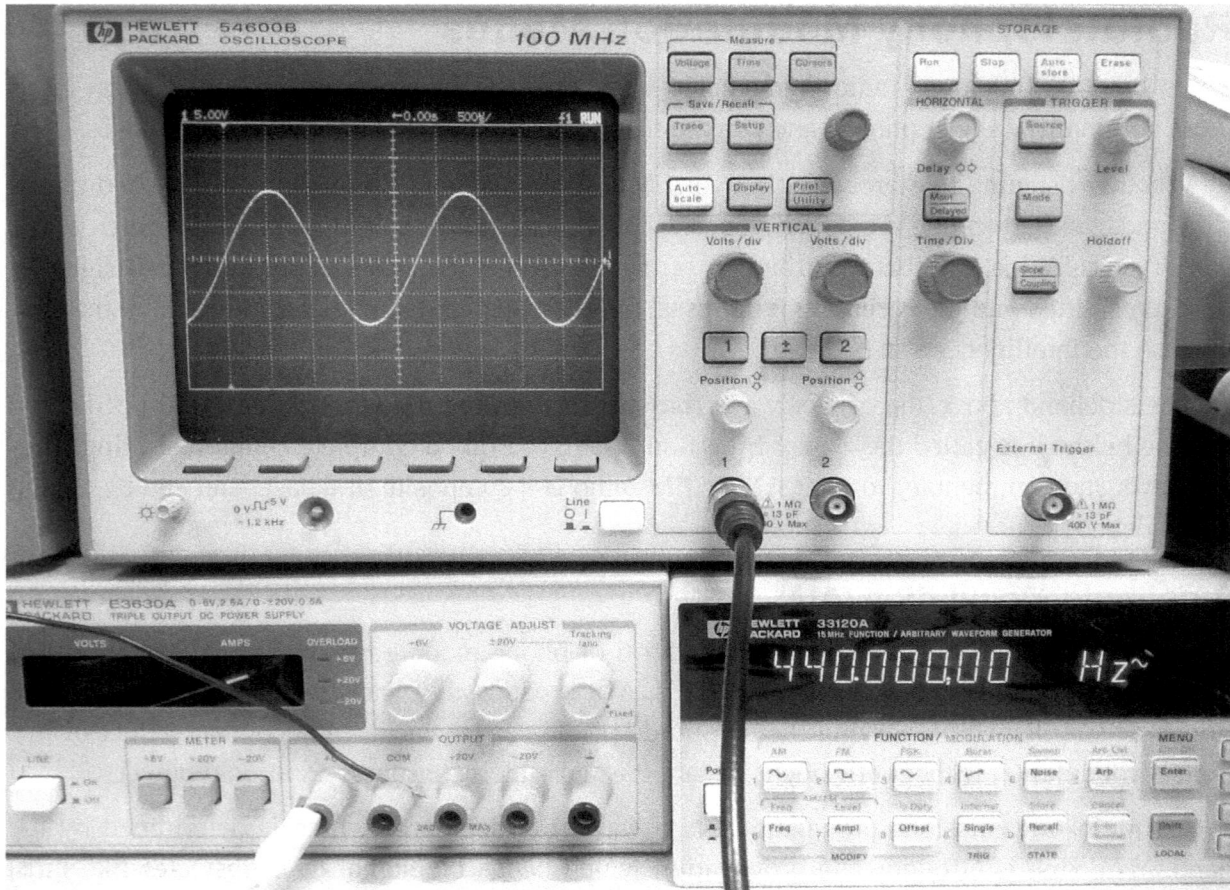

- **AM Sidebands:** The vertical height of the carrier varies with what is being transmitted; the up and down pattern looks similar to a roller coaster, and the sidebands expand and contract the bandwidth as the amplitude varies.

- **FM Sidebands:** The horizontal length of the of the signal varies with what is being transmitted. Ergo, the carrier band waves are stretched and compressed like an accordion as shown on an oscilloscope.

Now that you understand the similarities between AM and FM mode, let's throw the book away as they say, to learn about a third mode called single sideband modulation(SSB), one that will become an essential part of your community communications strategy.

Single Sideband Modulation (SSB)

The first Single sideband modulation (SSB) patent was awarded on December 1, 1915, to John Renshaw Carson. However, it was not until 1957 when the USAF Strategic Air Command established SSB as a standard for its aircraft that this modulation began receiving broad popular support from radio manufacturers and amateurs alike. This is because when it comes to range and energy efficiency, it runs rings around the AM and FM modes.

With SSB, the transmitter removes the carrier wave (baseband) and either the upper or the lower sideband. Then the intelligence is concentrated on the remaining side band, selected by the operator.

With SSB, receiving operators must know both frequency used for the transmission and the sideband as well- SSB USB (upper) or SSB LSB (lower).

For long distance (DX) contacts, AM is much less efficient than SSB and should only be used in special circumstances; for example, if the radio in the receiving station does not support SSB mode as in CB.

That being said, AM, FM and SSB all have their place and this older analog technology is tried and true. However, there is a new modern modulation that offers high quality digital voice under poor band conditions.

Digital Voice Modulation (DV):

Unlike AM, FM and SSB, which are analog radiotelephony modulations, digital voice modulation (DV) uses binary code digital signal. DV signals are created when an analog audio signal is digitally encoded and transmitted as binary ones and zeroes. On the receiving side, the binary ones and zeros are decoded back into an analog audio signal.

The modern modulation has attractive benefits. It uses the least amount of bandwidth, reduces clutter, and the sound is clean and free of hiss. Nonetheless, the DV mode is highly dependent on specialized technology.

DV is still in the early adopter phase and will eventually go mainstream with amateur HAM operators. While a technology like DV in its early adopter phase is enticing, you need to focus on the technologies that are clearly in their mainstream acceptance phases.

As the leader of a survival community and as the founder of a local Radio Free Earth network, the important thing for you to remember is this: CW is the most efficient way to send Morse code, and SSB is the most efficient way to communicate by phone (voice) over the greatest possible distances and with the least amount of transmit power.

Still the same, some may wonder why one should buy more expensive amateur HAM radios when less expensive CB and GMRS radios are available. The answer is amateur HAM radios are more robust and offer more modes, such as RTTY and Digital Data Modes, and they allow very long range communications, sometimes using very low transmit power.

RTTY and Digital Data Modes

Unlike digital voice modulation (DV), various digital data modes have been used for decades and are well-established. Like DV, these digital data modes require radios to be interfaced with computers, sound cards, external devices and advanced software.

There are many different digital data modes, but the following three are among the most popular and will certainly be useful for survival communications. The oldest of these data modes is RTTY.

- **Radio teletype (RTTY):** This mode dates back to the use of hard-wired teleprinters (teletype machines) that used the 5-bit Baudot code instead of Morse code.

  HAMs use personal computers with RTTY software to emulate early teleprinters. This time-tested communication mode supports real-time keyboard-to-keyboard chats. RTTY throughput is a typing speed of 60 to 100 words per minute depending on RTTY mode and operators' typing skills.

- **Packet:** This digital mode predates the internet and is often called pmail because messages are sent, and then received and stored in a manner similar to Internet email. It requires a device called a terminal node controller (TNC) which connects your transceiver with a personal computer.

  This mode was the first used with a networking scheme called Digipeaters. These stations listen for messages, record them in temporary storage and then re-transmit them. This allows pmail messages to be automatically relayed by a series of Packet Node stations to a designated recipient station.

  This digital data mode is a real time saver because it offers unattended automation and error checking. You can reliably exchange large volumes of pmail in an unattended manner between the base stations of friendly communities during off hours.

- **Digital Data Modes:** There are several excellent digital data modes available today, and PSK31 (Phase Shift Keying, 31 Baud) is currently the most popular among amateur radio operators. With this mode, the radio is connected to a personal computer with a reasonable quality sound card. It lets operators conduct real time keyboard-to-keyboard chats at typing speed of about 50 words per minute, or text messages may be easily sent.

  The drawback with PSK31 is that unlike the Packet mode, there is no built-in error checking. With some digital software, accessory programs are available which do offer error correction for the PSK modes. And there are other digital data modes that offer standard error checking and other advantages including image transfer.

Computer software is required with digital data modes, and two freeware programs are worth mentioning here:

Packet Manager Software: A terminal node controller (TNC) is required for packet mode operations and can be controlled directly with a dumb terminal using software built in to the TNC.

The popular packet program, Outpost Packet Message Manager, offers a more intuitive graphical interface that helps operators to organize their pmail in similar fashion to popular email programs. Outpost is free to the public. To learn more about it visit: www.outpostpm.org.

RTTY and Digital Data Modes Software: There are many paid and freeware programs available for running your transceiver in a digital data mode.

One of the most popular freeware HAM radio digital modem programs is called FLdigi. You can use it for CW, PSK31, RTTY and a wide range of other powerful digital modes, including CW. It can be downloaded for free, and there are a number of accessory programs that communicate directly with FLdigi.

There are different ways to send image files with various digital data modes, though some would argue, not as elegantly as with dedicated image modes.

Image Modes

As the old saying goes "a picture is worth a thousand words" and there are established ways to transmit these images over radio. Below is a partial list of those image modes most relevant to survival.

- **Amateur television (ATV):** These frequencies were assigned for use by HAM operators to transmit broadcast quality video and audio. It is also called HAM TV or fast-scan TV (FSTV). American HAMS use the National Television System Committee (NTSC) system ATV, which predates digital television. This image mode will be a good candidate for use after the global tribulation.

- **Slow scan television (SSTV):** A way to transmit good quality photographic images, monochrome or color. This will be especially useful way for field operators to transmit images of ground positions, hostile maneuvers and such. For base operations, this offers a way to share personal information such as family photos.

- **FAX480:** Send faxes via radio with higher resolution than SSTV. SSTV has a resolution of 240 lines, whereas FAX480 doubles that. For base operations, this image mode offers an excellent way to transmit technical information such as circuit board schematics.

- **Weather Facsimile (WEFAX):** A non-standard system that uses satellites and ground stations to transmit weather images and works in a similar way to SSTV. This national system will likely be the last to fall in the worst of times but until it does, it can provide life-saving information.

It is important to note that image modes typically require specialized equipment plus personal computers or laptops with high-fidelity sound cards, cameras, scanners and so forth.

Digital data and image modes need to be evaluated early on, as they require extra expertise, equipment and power. Therefore, before investing resources in any of these modes, first establish a clear requirement. If you have the need, then should come the questions about the extra technology and power required.

That being said, as the leader of a survival community and the founder of a local Radio Free Earth network, there is one mode you want to keep in mind for what I call the "backside."

Earth-Moon-Earth

The backside is that time at the end of the global tribulation when we see blue skies and taste sweet waters once again. This will be a critical moment in the history of our species, and many will wonder if humanity will slip into another dark age. Or can we begin a new renaissance with the promise of an enlightened Star Trek future?

This begs a very critical question. What can you plan for at the outset of the tribulation to occur on the backside that will galvanize hope for a Star Trek future?

There is one mode you want to keep up your sleeve so to speak, and in your thoughts. It is called Earth-Moon-Earth (EME) communication, and it is also known as "moonbounce." In theory, the concept is simple. A HAM operator uses the moon or even the ionized trail of a meteor as a passive reflector for transmissions which are then reflected back to the Earth.

Moonbounce is a powerful way to transmit messages because reflected EME signals can be received by EME receiving stations spanning a third of the surface of the Earth. Is this possible? Oh, yes, and HAMs are doing it today, so the technology and know-how you need are already available.

Now that we've covered the different modes and modulations used to transmit messages on a given frequency, the next step is to learn how these frequencies are organized into function-related spectrum regions, bands and sub-bands.

Chapter Review

The goal of this book is to give you the basic knowledge you need to create a vision and to be able to explain it to those who are going to have to implement that for you. They will do the job of hooking together the boxes and making things spark. Your job will be to create the vision and then to find the necessary people and funding to bring it to fruition.

Let's take a moment to quickly review the key points.

Radio Frequency Regulation

Government agencies create bandplan plans for the radio operators in their countries that specify who can access a given frequency and what they can do with it. Manufacturers design their radios according to these bandplan plans. Depending on the band, a license may not be required but there are universal rules for all bands that must be respected.

Modes and Modulation

A mode is how something works, and there are several modes. The most important one for survival is Morse code via CW, or continuous wave. For phone (voice communications) a modulation is required to create an outgoing transmission.

The four modulations commonly used for phone mode contacts as referenced in this book are AM, FM, SSB and Digital.

There are also digital data modes such as RTTY and Packet, and Digital Data modes such as the ever-popular PSK31. Image modes are also available such as amateur television (ATV) and FAX480. However, these modes require specialized equipment plus personal computers or laptops with high-fidelity sound cards, cameras, scanners and so forth.

Morse Code (CW)

Morse code was the only way to communicate for the first decade or so of wireless radio. It is based on landline telegraphy of the 19th century that include codes, abbreviations and procedures. Continuous wave (CW) is the established standard for highly-efficient Morse code contact. When all other modes and modulation fail, CW is the last, best option.

Radiotelephony Modulation Modes

Radiotelephony is commonly referred to as "Phone," and modulation describes the process in which the intelligence of your transmission (spoken voice, data, etc.) is combined with a carrier wave and sidebands.

- **Amplitude Modulation (AM):** AM is the oldest form of phone modulation. It is commonly used with CB radios, some of which can support SSB modes and with HF radios.

- **Frequency Modulation (FM):** Invented in 1933, frequency modulation (FM) is used for line-of-sight transmissions.

- **Single Sideband Modulation (SSB):** This modulation removes the carrier wave and one sideband and then concentrates the signal on the remaining sideband. In terms of broadcast power and range, it is more efficient than AM and FM, therefore is commonly used for DX contacts.

- **Digital Voice Modulation (DV):** DV uses a computer-like binary code digital signal. It offers a cleaner sound and is more efficient with bandwidth. However, universal amateur standards are still in flux, so commercial radios are recommended for this modulation.

Earth-Moon-Earth (EME)

Commonly called "moonbounce." EME offers a way to bounce a transmission off the surface of the moon that returns to cover a third of the face of the Earth. Keep it in mind for the back-side (when we see blue skies and taste sweet waters once again) for morale building.

In the next chapter, we'll see how the FCC Frequency Band plan determines who can transmit on the various frequencies licensed for amateur use and which modes are available with those frequencies and applicable HAM license classes.

3

Frequency Bandplan

In the previous chapter, we examined the wide range of functionality available with modes and modulations. Due to government regulations, some modes and modulations are specifically allowed or disallowed for various frequencies. Therefore, this chapter presents a building brick approach to describe how radio spectrum regions, frequencies, bands, and sub-bands are organized with what is known as the FCC Frequency Bandplan.

What makes the FCC bandplan confusing for beginners is that access to any given frequency and the modes, and modulations operators have available for use will depend on their license. The result is that an FCC bandplan can look as though someone threw a pot of spaghetti on the wall and what stuck, stuck. Add to that, there are other factors that further increase the apparent jumble.

One such jumble factor is how radios are made. To visualize this in your mind, let's imagine we're shopping in a thrift store for antiques and collectibles, and we come across an old analog shortwave radio receiver with manual tuning made in the 1960's.

The radio tuner dial offers three different shortwave bands: Shortwave 1, Shortwave 2 and Shortwave 3. Are these three shortwave bands designated by the FCC? No. In fact, they never were nor never will be. So why are they shown on the tuner in this thrift shop radio?

The Shortwave 1, Shortwave 2 and Shortwave 3 band choices shown on our old thrift shop radio were created by the manufacturer to make it easier for consumers to use their radios. What the manufacturer did in this example was to divide the High Frequency (HF) radio spectrum region into three arbitrary shortwave bands.

Here is one of the oddities one needs to remember about HF. When communicating with a two-way radio, HF is the proper designation. However, when HF frequencies are only being used for listening, to broadcast and for two-way radio traffic, shortwave is the proper designation.

For amateur HAM operators, the FCC bandplan for the HF radio spectrum region encompasses several bands, sub-bands and frequencies. What licensed amateur HAMs can and cannot do within the HF radio spectrum region will depend upon the type of license they hold. With this in mind, let's take a quick look at HAM licenses.

HAM Licenses

To become a licensed HAM radio operator, you must take and pass one or more written examinations before you are issued a license. As was pointed out earlier, a demonstration of Morse code proficiency is no longer required to obtain a license, but the ability to demonstrate an understanding of the technology at each license level and FCC regulations is necessary.

At present, there are three licenses offered to newcomers. Technician, General and Extra. For the sake of simplicity, a quick way to understand the major differences is as follows:

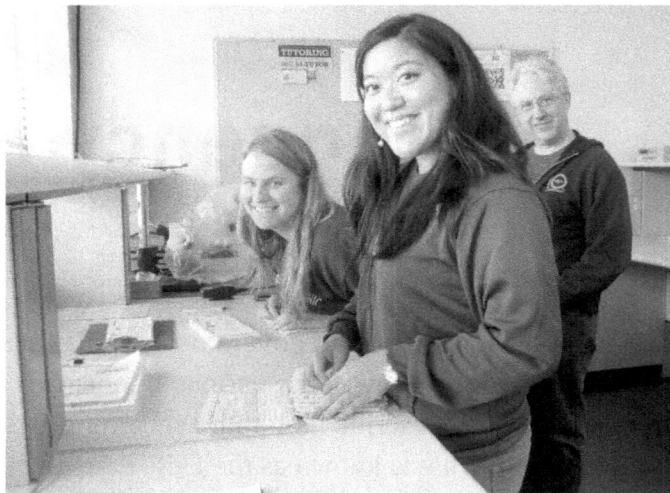

License	USA Contacts	Intl. Contacts	HAM Frequencies
Technician	Yes	No	Some
General	Yes	Yes	Most
Extra	Yes	Yes	All

If you are starting from scratch, you must first pass the Technician examination. This license will come with a permanent assignment of a call sign, and you are allowed to make contacts within the United States but not international contacts.

For this reason, after passing the Technician examination it will behoove you to study to take the General license test. When you pass it, you will have the right to make international contacts, and you will have access to most all of the frequencies designated for use by amateur HAM radio operators.

For those who want everything and who hold a General license, there is the Amateur Extra rating and this is the most difficult test of the three to take. While it is a difficult exam, the additional frequencies allowed with this rating, make it well worth the effort to obtain.

The operative word here is "frequencies" because frequencies are fixed.

Wavelengths and Hertz

No matter where a frequency is located on the radio frequency spectrum, it will always be identified by the same two characteristics: wavelength and cycles per second (Hertz.)

Wavelength

Radio signals travel in electromagnetic waves. The wave begins at a zero point which is in the center, then it ascends to the uppermost limit. From this point, it descends past the zero point at the center to its lowermost limit. When it ascends back to the zero point, the wave is complete. Ergo, one cycle of this entire up, down and up again wave motion is called the wavelength.

To illustrate the point, let's use a frequency with a wavelength of 6 Meters or "6 M" as you'll see it in radio advertisements. In this case, the wavelength is 6 Meters which means that for the signal to complete the entire up, down and up again wave motion, requires a wavelength of 6 Meters (19.685 Feet).

What this means is that in order to broadcast on this frequency with a full wavelength, you'll need an antenna that is 6 meters (19.685 feet) long. More commonly used antennas are ½ or ¼ wavelength.

Now that we've defined length, let's tackle the second attribute which brings us to an old electrical term, cycles per second.

The next question is, how many 6 Meter-long wavelengths travel the length of the antenna each second.

Hertz (Cycles Per Second)

Previously, we established that a single wavelength includes one full cycle where there a complete up, down and up again wave motion. How often this wave traverses the length of the antenna each second is measured in Hertz, named for Heinrich Rudolf Hertz, an early electromagnetic researcher. The actual term used in the early days of radio was cycles per second, which is more descriptive.

To illustrate the concept of Hertz (cycles per second), let's return to the 6 M wavelength example from above. The first or lowest frequency with an Amateur 6 Meter wavelength is 50.00 MHz. This means that 50 million complete 6 Meter wavelengths will radiate off our full wave antenna each second.

Given than 50.00 MHz is the first frequency with a 6 Meter wavelength, what is the last? In this case, the last frequency of a 6 Meter wavelength is 54.00 MHz. This means that 54 million complete 6 Meter wavelengths will radiate off our full wave antenna each second.

Now, let's go from a 6 Meter wavelength to a 17 Meter wavelength. With the 17 Meter band, the first frequency is 18.068 MHz. This means that 18.068 million complete 17 Meter wavelengths will radiate off a full wave antenna each second. Also, note that the last 17 Meter frequency is 18.168 MHz.

When we compare the span of frequencies between the 6 Meter wavelength and the 17 Meter wavelength we see a profound difference. With 6 Meter, there is a frequency difference of 4.0 MHz between first and last, whereas with 17 Meter, the difference is only 0.10 MHz.

When we compare the examples above, we see that the first and last frequencies of wavelength do not follow a predictable pattern, because they are assigned with the FCC bandplan. Another item of note is how the frequencies are represented in Hertz. In our two examples, we used the 6 Meter wavelength with a starting frequency of 54.00 MHz and the 17 Meter wavelength with a starting frequency of 18.068 MHz.

Given that we've used the abbreviation MHz several times in our example, what exactly does it mean and where does it fit withing the grand scheme of things? The table below explains Hertz terms commonly used with amateur radio.

Term	Abbr.	Description	Amateur Radio Use
Hertz	Hz	1 Hertz (Hz) means that something happens once a second.	Hertz is a reference term only as the radio spectrum is 3 Hz to 3000 GHz.
Kilohertz	kHz	1 Kilohertz (kHz) means that something happens one thousand times a second.	A few long wavelength frequencies are available for General and Extra licenses.
Megahertz	MHz	1 Megahertz (MHz) means that something happens one million times a second.	Most all amateur HAM frequencies are in the MHz range.
Gigahertz	GHz	1 Gigahertz (GHz) means that something happens one billion times a second.	A limited number of short wavelength frequencies used for special purposes such as amateur satellite.

What the table above shows is that frequencies in the Megahertz (MHz) range will dominate those you will use each day. Likewise, there is a predictable inverse relationship between wavelength and frequency that you need to mindful of:

- As wavelength increases, frequency decreases.

- As frequency increases, wavelength decreases.

Now we're ready to take this up a notch by looking at how frequencies are organized into radio spectrum regions, bands and sub-bands.

Spectrum, Regions, Bands and Sub-bands

One of the difficult things about understanding how frequencies are organized is the confusion of definitions often encountered on the Internet. It is one of those situations where HAMs come to understand the confusion. This way, even when folks get it wrong, they'll still understand the context. However, for newbies this a grueling learning curve issue.

With this in mind, I chose one of several ways to explain the organization which is based on a four-level hierarchy where the radio spectrum is the complete radio frequency spectrum, which is then divided into radio spectrum regions, bands and sub-bands. The goal is to help the reader better understand radio basics when evaluating their communications needs.

The complete radio frequency spectrum ranges from 3 Hz to 3000 GHz and to enable a comprehensive bandplan for all nations, the International Telecommunication Union (ITU) has divided the radio spectrum it up into twelve separate regions with the following hierarchy:

- **Regions:** A region is a contiguous group of radio spectrum frequencies organized into bands according to general characteristics. The lowest region is Extremely low frequency (ELF) and the highest is Tremendously high frequency (THz) or Terahertz range. A region can have many bands.

- **Bands:** Contiguous frequencies organized by purpose or intended use. A band has a beginning and ending frequency. A band can have many sub-bands.

- **Sub-bands:** Small clusters of frequencies within a band. Limited frequency ranges within a band are used to designate access based on license privileges and allowable modes. Each sub-band has a beginning and ending frequency. Sub-band frequencies cannot exceed the beginning and ending band frequencies.

Common sub-band modes in the FCC bandplan include:

- RTTY and Data
- Phone and Image
- CW Only
- SSB Phone (SSB or USB)
- USB Phone (USB Only), CW, RTTY and Data
- Fixed Digital Message Forwarding Systems.

As the leader of a survival community and as the founder of a local Radio Free Earth network, the three radio spectrum regions you will use extensively are high frequency (HF), very high frequency (VHF) and ultra high frequency (UHF).

High Frequency (HF) – 3 to 30 MHz

Of the three radio spectrum regions, high frequency (HF), aka shortwave, offers you the greatest distance, whether it intrastate, interstate, regional or international. This is because of the two ways in which HF signals propagate:

- **Groundwave:** A method of HF propagation that uses the area between the surface of the earth and the ionosphere for transmission. Radio waves follow the contour of the earth and can propagate for considerable distances.

- **Skywave:** The other form of HF propagation is called skywave and is commonly referred to as "skip" because transmitted radio waves reach the ionosphere, an electrically charged layer of the upper atmosphere, and then are reflected or refracted back toward towards the ground.

Other factors will have a direct bearing upon your ability to make long distance (DX) contacts with HF, such as sun spot activity and the antenna used. However, when compared with line-of-sight bands, a properly configured HF transceiver will always give you the longest range.

Wavelength	Base Frequency	FCC Assigned
80 Meters	3.5 MHz	3.500 – 4.000 MHz
60 Meters	5.3 MHz	5.332, 5.348, 5.358.5 5.373 & 5.405 MHz
40 Meters	7 MHz	7.0 – 7.3 MHz
30 Meters	10.1 MHz	10.1 – 10.15 MHz
20 Meters	14 MHz	14.0 – 14.35 MHz
17 Meters	18 MHz	18.068 – 18.168 MHz
15 Meters	21 MHz	21.0 – 21.45 MHz
12 Meters	24 MHz	24.89 – 24.99 MHz
11 Meters	26 MHz	CB Channels 1 – 40
10 Meters	28 MHz	28 – 29.7 MHz

Of particular note are three HF bands: 40 Meters, 20 Meters and 11 Meters. These bands will be the first ones you'll want to evaluate for your community communication strategy.

- **40 Meters:** Local (Intrastate) area coverage during the day. International DX at night.

- **20 Meters:** Good coverage during the day and night. The is most reliable band for DX.

- **11 Meters:** This is the Citizens Band. Here is an interesting aspect of DXing, as it is called. A transmission from CB radio running in its standard AM mode can be heard from 1 to 25 miles away depending on equipment and conditions.

However, switch that same CB radio into SSB mode, and you can start making DX contacts in Canada, Mexico or even Europe. Again, this depends on equipment and conditions.

When evaluating all available HF frequencies, here is a rough rule of thumb:

- 20 Meters (14 MHz) is a popular mid-range that works to some extent day or night.

- Lower frequency bands above 20 Meters generally work better at night and include: 30 Meters (10.1 MHz), 40 Meters (7 MHz) and 80 Meters (3.5 MHz).

- Higher frequency bands below 20 Meters generally work better during the day and include: 17 Meters (18 MHz), 15 Meters (21 MHz), 12 Meters (24 MHz) and 10 Meters (28 MHz).

- The farther away you get from 20 Meters in either direction, the more pronounced these effects become.

Again, these suggestions are a rough rule of thumb because during the global tribulation the effects of space weather and other considerations will be more pronounced.

Very High Frequency (VHF) – 30 to 300 MHz

VHF is a line-of-sight spectrum region that is well-suited to operations in the field, away from man-made structures. The most popular band in this region is 2 meters. When you see an amateur HAM radio operator carrying a handy-talkie (HT), it will likely support the 2 Meter band in the VHF radio spectrum region.

Wavelength	Base Frequency	Available Frequencies
6 Meters	50 MHz	50 – 54 MHz
2 Meters	144 MHz	144 – 148 MHz
1.25 Meters	222 MHz	222 – 225 MHz

Two VHF bands are oddballs of a sort. They are 6 Meters and 1.25 Meters. The 6 Meter band is cross-over band from VHF to HF-like characteristics of the 6 Meters band has inspired amateur operators to dub it the "magic band" though it is not heavily used.

The low popularity of the 1.25 Meters band is more a case of FCC mismanagement than anything else. Few transceivers support the 1.25 Meters band so its only virtue is that there is not a lot of traffic on this band.

Ultra High Frequency (UHF) – 300 to 3000 MHz

UHF is a line-of-sight spectrum region that is well suited to operations in man-made structures, due to short wavelengths. Consequently, UHF signals can more readily pass through the construction materials in man-made structures, especially the rebar in reinforced concrete.

Wavelength	Base Frequency	Available Frequencies
70 Centimeters	420 MHz	420 – 450 MHz
65 Centimeters	462 MHz	FRS/GMRS Channels

Of note is the 65 Centimeters band. This is not a HAM band, but it does give you an approximate location of FRS/GMRS Channels on the radio spectrum.

Like the CB band, transmitting on frequencies assigned to low-power Family Radio Service (FRS), it does not require a license, though the power limitations of FRS make it marginally useful at best.

On the other hand, the GMRS Channels do require a license but not a proficiency examination. Depending on the channel:

- FRS allows for transmit power of between 0.5 Watts to 2.0 Watts.
- GMRS allows for transmit power of between 0.5 Watts to 50.0 Watts.

Obviously, inexpensive GMRS radios offer a viable line-of-sight alternative to more expensive amateur HAM radios for the budget-minded. For example, walkie-talkie GMRS radios offer transmission power levels comparable with licensed HAM UHF Handy-Talkies (HT).

Chapter Review

This chapter has presented you with a high-level view of the key concepts and terms you'll need to effectively organize a community communications strategy with your engineering team and to make basic choices about the types of equipment you will purchase in support of your goals.

Let's take a moment to quickly review the key points.

Radio Spectrum Regions, Frequencies, Bands and Sub-Bands

A frequency represents a fixed and immovable spot in the radio in the radio frequency (RF) portion of the electromagnetic spectrum.

- **Regions:** There are several spectrum regions. Three of these, HF, VHF and UHF, will be the workhorses of a survival community and a region can have many bands.
- **Bands:** Frequencies organized by a common purpose with a beginning and ending frequency. A band can have many sub-bands.

- **Sub-bands:** Small clusters of frequencies within a band used to designate access based on license privileges and allowable modes.

Remember, frequencies are fixed. Conversely, regions, bands and sub-bands are arbitrary groupings organized to serve a regulatory purpose.

HAM Licenses

The FCC currently issues three kinds of amateur HAM radio licenses. Technician, General and Extra. The Technician rating is the entry level and comes with a permanent assignment of a Call Sign. The two ratings above Technician are General and Amateur Extra.

Wavelengths and Hertz

No matter where a frequency is located on the radio frequency spectrum, it will always be identified by the same two characteristics: wavelength and cycles per second (Hertz.)

- **Wavelength:** Radio signals travel in waves. The wave begins at a zero point which is in the center, then it ascends to the uppermost limit. From this point, it descends past the zero point at the center to its lowermost limit.

- **Hertz (Cycles Per Second):** How often this wave traverses the length of the antenna each second is measured in Hertz. The actual term used in the early days of radio was cycle per second, which is more descriptive.

There is an inverse relationship between wavelength and frequency. As wavelength increases, frequency decreases, and as frequency increases, wavelength decreases.

Regions, Bands and Sub-bands

A frequency is a fixed value in the radio frequency spectrum. Regions, bands and sub-bands are arbitrary man-made groupings.

- **Regions:** Contiguous frequencies organized into bands according to general characteristics within a region. A region can have many bands.

- **Bands:** Contiguous frequencies organized by purpose or intended use. A band can have many sub-bands.

- **Sub-bands:** Small clusters of frequencies within a band. Generally used to designate access based on license privileges and allowable modes.

The three radio spectrum regions you will use with your community communications plan are:

- **High Frequency (HF) – 3 to 30 MHz:** HF aka shortwave, offers the greatest distance, whether it's intrastate, interstate, regional or international.

- **Very High Frequency (VHF) – 30 to 300 MHz:** A line-of-sight spectrum region that is well suited to operations in the field, away from man-made structures.

- **Ultra High Frequency (UHF) – 300 to 3000 MHz:** A line-of-sight spectrum region that is well-suited to operations in man-made structures, due to short wavelengths.

Two bands that are not typically used by licensed HAMs are Citizens Band (CB) and FRS/GMRS. These bands are limited in transmission power and range. However, these less robust radios can get you on the air without breaking the bank. So let's go shopping.

4

Radio Shopping Basics

After reading the last two chapters, one of three things might be going through your mind:

- **Old Hands:** If you're already an amateur HAM radio operator, by now you should be seeing how this book will help newbies with checkbooks. If so, and you're saying this approach works, (even if you might have explained some things differently) all I can say is, the first round is on me.

- **Newbies:** If you feel like you're doing the backstroke in a lap pool filled with ice cold alphabet soup, welcome to the club. This is a rite of passage for everyone, and it gets easier as you work with the technology. If this is you, then you'll appreciate this and the following chapters in the book because related concepts will be re-summarized as needed. No external or internal references should be expected.

- **All Others:** Thanks for sticking with it. Now we start to have some fun.

Before you go shopping for a radio, you need to know what you're looking for. For example, a test drive of a 400 horsepower sports car may be a real hoot but if you're looking for a minivan for the family, a hot rod is not what you need to test drive.

With HAM radios, there are the plain Jane minivans and the sexy sports cars as well. Again, practical outweighs sexy, and practical criteria that we'll review in this chapter will help you clearly define your needs and goals to maximize the survival utility of your radio purchases:

- **Radio Services:** The FCC uses services to categorize radios based on who will use them and for what applications. This is the broadest radio shopping criteria.

- **Watts and Bands:** Each type of radio purchased must serve a communication mission defined by the radio's transmit power and supported bands, frequencies or channels.

- **Volts and Amps:** Power is the most decisive logistical factor in support of a radio mission. Specifically, what types of power sources will be available and how much power will a transceiver draw for standby, receive and transmit modes.

- **DC Power Management:** Today's modern radios are powered by AC and DC sources. For survival mission roles, only DC powered radios are suitable; and do not be surprised if you spend more on a radio's power source than on the radio itself.

In our example above, we compared a sports car with a minivan, two very different vehicles which can easily be distinguished from a distance. With two-way radios, this is not so because of all the different makes and models of two-way radios on the market today. Consequently, while two different radios can look very similar on the outside, what they enable you to do on the inside can be as different as night and day.

Therefore, to help you begin aiming your efforts efficiently, a brief discussion of radio services is necessary.

Radio Services

Radios tend to look alike to newbies, so the easiest way to distinguish between them is to ascertain the FCC service applications they are designed for. For example, the Business Radio Service is intended for business users and the Amateur Radio Service is reserved for HAM operators. To avoid granularity, we'll use three broad categories in our discussion of these various services:

- **Commercial:** Radios in this category are made to serve government and business needs with the highest levels of performance and reliability. To paraphrase that old Timex advertising slogan, Commercial radios are built to *"take a licking and keep on ticking."* Consequently, they are the most expensive.

  If you are interested in a commercial service grade radio, it will likely be for a limited survival role such as long range ground reconnaissance or FM broadcasting, and you'll need expert advice and a hefty budget. For this reason, we'll focus on amateur and consumer service radios.

  Quality Reference Brands: Motorola (USA) and ICOM (Japan)

- **Amateur:** These radios are the mainstay of amateur HAM radio operators and will become the workhorses of your overall community communications strategy. Versatility, efficiency and long distance communications are the strengths of these radios.

  Top Three Brands: Kenwood (Japan), Yaesu (Japan) and ICOM (Japan).

Value/Specialty Brands: Alinco (Japan), Elecraft (USA).

Low-cost "Throw Away" Brands: BaoFeng (China) and Wouxun (China)

● **Consumer:** Most frequently called "bubble pack radios" these radios are simple to operate and widely available. The two variants you'll employ in your community communications strategy will be Citizens Band (CB) and General Mobile Radio Service (GMRS).

Quality CB Reference Brands: Cobra (USA), Midland (USA) and Uniden (Japan)

Quality GMRS Reference Brands: Motorola (USA) and ICOM (Japan)

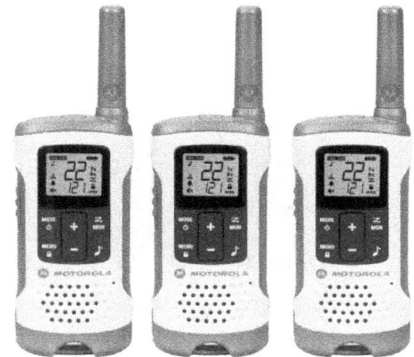

The big difference between commercial, amateur and consumer radios is frequencies versus channels. With amateur radios, frequencies are assigned to tuneable bands which offer vast array of different frequencies depending on the operator's regional bandplan and license class. With commercial and consumer radios, frequencies are assigned to a relatively small number of channels by comparison.

A general rule of thumb is that amateur radios offer more frequencies but are more complex. Commercial and consumer radios are ideal for low-skill users. While they offer much fewer channels (each with an assigned frequency) they are far simpler for newbies to operate than modern amateur radios.

Watts and Bands

A well-balanced community communication strategy will provide suitable radios for each designated communication mission. These mission requirements will, in turn, be driven by the radio's transmit power on the bands, frequencies or channels it serves.

For example, when shopping for radios, one of the most popular types purchased by amateur HAM beginners are dual-band FM Handy-Talkies (HT). A common thing to see with advertised model descriptions is something like 5-watt Dual Band 2M/70cm. Let's break this description down into its three constituent parts: transmit power, band(s) and frequencies.

Transmit Power

In our example description we see "5-watt." The term watt (abbreviated W) is used to describe the rate at which electromagnetic energy is radiated. As a rule, the smaller the radio, the fewer the watts it will transmit.

Our example HT radio is a small, handheld amateur HAM transceiver that transmits a signal with 5 watts of power. For a larger mobile transceiver installed in a car, the transmit power is usually 25 to 50 watts. Then, in the home base station, a HAM operator with full-size transceivers can transmit signals with up to 1500 watts on most HF frequencies.

Bands

As we saw in the previous chapter, there are many different bands allocated for use by licensed HAM radio operators. In our example, the handheld transceiver is advertised as a "Dual Band" radio which makes it a multi-band transceiver. Which of the following three descriptions identifies this example radio?

- **Mono-band:** The radio only supports one band as defined by wavelength or range of frequencies. If our example radio were a mono-band, the advertisement could read as 5-watt 2M.

- **Multi-band:** Our example is a multi-band radio and most handheld radios are dual-band types; however, a multi-band radio is not limited to two bands. Let's assume our example radio also handled the 1.5 M band. The advertised description could be something like: 5-watt Three Band 2M/1.5M/70cm.

- **All-band:** Some transceivers will support all of the bands allocated for use by amateur HAM operators in all of the frequency spectrum regions. However, the term all band can also be limited to a range of bands within a given spectrum region such as high frequency HF. In this case, the advertisement will note the limitation with something such as: HF All Band.

Regardless of how many bands a radio will support, these bands can be expressed using their wavelength or the allocated base frequency for the band. Returning to our example radio, the description is 2M/70cm, where 2M is the wavelength of the 144MHz to 148MHz portion of the VHF band and 70cm is the wavelength of the 420MHz to 450MHz portion of the UHF band.

However, this same radio could be described in three other ways.

- **144/420:** 144 is the base frequency for the VHF band and 420 is the base frequency for the UHF band.

- **2m/420:** 2M is the wavelength of the VHF band and 420 is the base frequency for the UHF band.

- **144/70cm:** 144 is the base frequency for the VHF band and 70cm is the wavelength of the UHF band.

The point here is that all variants shown above are correct and the fact is manufacturers like to mix the descriptions. Yet, as the old saying goes, "a rose by any other name is still a rose." That is, unless you see something odd that does not follow this simple convention.

Chinese Radios and Frequencies

One such oddball transceiver that does not follow this simple convention is the dual band Chinese BaoFeng BF-F8HP. A handy talkie radio, it is advertised with the following frequency ranges: 136-174MHz VHF & 400-520MHz UHF.

At first glance, one could assume that 136-174MHz VHF & 400-520MHz UHF applies to receive mode (RX) only and that a narrower range of frequencies is available for transmit (TX) mode. However, if you read the specification for a Chinese radio such as this, you'll see that the 136-174MHz VHF & 400-520MHz UHF applies to both RX and TX modes.

On the other hand, transceivers manufactured for the American market by firms in America, Japan and elsewhere adhere faithfully to the allocated band plan. While these radios receive on a wider range of frequencies, they are limited to transmitting on a more narrow range of frequencies as allocated by international and national governing bodies. Therefore, the range of transmit frequencies is dramatically different.

- **Chinese TX:** 136-74MHz VHF & 400-520MHz UHF
- **American TX:** 144-148MHz 420 to 450MHz UHF

Given the two are so different, what is happening here? It is a combination of regulation and manufacturer choices.

The highest governing authority for management of the global radio spectrum in the world is the The International Telecommunication Union (ITU) and national agencies such as the FCC here in the USA work within the allocations assigned by the ITU.

To address regional and national needs, the ITU has divided the world into three regions.

- **Region 1:** Europe, Africa, the former Soviet Union, Mongolia, and the Middle East, west of the Persian Gulf, including Iraq.
- **Region 2:** The Americas including Greenland, and some of the eastern Pacific Islands.
- **Region 3:** Most post-Soviet states, Asia east of and including Iran, and most of Oceania.

Here is where the differences come into play between Chinese manufacturers and those of other countries. The Chinese offer flexibility so that one radio design can be used in many different regions. The programming required uses software which allows the selection of the region the radio will be operated in which in turn permits selection of frequencies allowed in those regions.

Does this mean that if a Chinese radio gives you the ability to transmit on a frequency that has not been allocated to your region and country, that you can do so legally? No, and if the FCC shows up at your door because you've been transmitting on an unassigned frequency, telling them to contact the manufacturer will get you nowhere.

With this in mind, here is the bottom line with Chinese radios and the frequencies they offer:

- **Know Your Country's Bandplan:** Regardless of what radio or brand you choose, you need to understand your bandplan and what you can do given the class of license you hold: technician, general or extra.

- **Double Check RX and TX Frequencies:** When purchasing any radio, always check the RX and TX frequency ranges. With RX, if you can hear, you're good. With TX, if you are on an unassigned frequency or using the wrong mode, you can get into a bind.

- **Double Check TX Transmit Power:** Access to an allocated frequency is regulated with a combination of license class, mode and transmit power. Always make sure of the transmit power settings available with any radio as most typically offer a range of low to high power settings.

  Remember, using more TX power than allowed is illegal. One more thing needs to be said about transmit power. Use only the amount of power necessary to make the contact. If 10 watts gives you reliable communications, do not use 100 watts as this will likely cause interference to someone else using the frequency.

A final note on Chinese radios and the ITU. In the absence of regulatory agencies during the global tribulation, Chinese radios that transmit outside of the ITU and national band plans for your area do offer added frequency options. However, these added frequency options do not constitute a significant purchasing reason and furthermore, can get you into a bind.

Again, I must stress the following. Just because a radio gives the ability to do something, "you have a right to do it" is not a legal defense. Ignorance of the law is no excuse.

Now that we've discussed the power your radio is using to transmit your signal into the radio spectrum, let's turn our attention to the power coming into the radio to make that possible.

Volts and Amps

While watts and bands define the communication mission served by each radio, volts and amps define the most decisive logistical factor in support of that mission. In other words, without ammunition, an expensive rifle becomes a club or a door stop. The same applies to the relationship between radios and the power needed to operate them.

In this chapter, we will use a simple power table for a generic example for a mobile radio.

Example Mobile Radio			
Input Power (VDC)	**Current Drain Amps (A) / Milliamps (mA)**		**Transmit (Watts)**
VDC Input Range	**Standby Drain**	**Max Transmit Drain**	**Maximum Output**
12 to 13.8 V DC	1.5 A	20.0 A	50 Watts

In terms of radio purchasing, the terms, volts, amps and watts, are consistently applied, and while a technical understanding of differences is helpful, what you need to remember is where these terms apply in the process of operating a radio.

Input Power (VDC)

A radio is no different than a hair dryer. You need a power source to make it work, and this is rated in volts. However, the difference with input power here is between alternating current (AC) and direct current (DC) and the proper operating voltage.

A hair dryer uses an AC power source rated in high voltage whereas your community communication strategy will be based on radios that use DC power, which require DC power sources rated as low voltage DC (VDC). Ergo, a volt is a volt whether it is DC or AC.

In the example above, a DC range is shown. This is because most modern radios can be powered with a range of volts. In the example above, the range is 12 to 13.8 V. However, manufacturers will specify this range in other ways such as DC 13.8 V ±15%, which translates to 11.73 V to 15.87 V.

Either way, we see that this example radio can use a 12 V power supply such as a car battery. So where does the 13.8 V specification come from? Car batteries.

While the battery you buy for your car is rated at 12 V, car batteries typically put out 12.6 V. Then, when your alternator is charging your battery, it puts out 13.8 V for a battery that is nearly charged. If the battery is heavily drained, the alternator in your car will put out as much as 14.2 V to recharge it.

Ergo, a mobile radio that is rated for an input source of 12 to 13.8 V is designed to handle the power output of a charged 12 V battery as well as the power output of an alternator as it charges the battery.

So then, how does this apply to our example mobile radio? As the term 'mobile' implies, the radio is intended for use in a vehicle. When you connect your radio to the battery in your car it works without the need for an intermediate power conversion device. Likewise, the same holds true if you're using a mobile radio with a backpack setup. In this case, you'll be using smaller 12 V gel cell batteries such as those used to power emergency lighting or a fish finder, for example.

Regardless of the radio or power source, understanding the amount of power your radio will drain is essential.

Current Drain Amps (A) / Milliamps (mA)

In the example mobile radio table shown below, the column heading Current Drain Amps (A) / Milliamps (mA) has two sub-columns.   Standby Drain which shows 1.5 A and Max Transmit Drain which shows 20.0 A. So what does this mean?

Example Mobile Radio			
Input Power (VDC)	**Current Drain Amps (A) / Milliamps (mA)**		**Transmit (Watts)**
VDC Input Range	**Standby Drain**	**Max Transmit Drain**	**Maximum Output**
12 to 13.8 V DC	1.5 A	20.0 A	50 Watts

The term amps stands for ampere with the abbreviation "A" and the term milliamps stands for milliampere and uses the abbreviation "mA". One mA is equal to one thousandth of an ampere. For example, 200 mA equals 0.2 A. It is important to remember this distinction because you will be buying radios with drain specifications rated in both A and mA.

However, in simpler terms, let's imagine that someone hands you a full 12 ounce glass of water and calls it 12 volts. How fast you drain the water from the glass is described in amps (A) and milliamps (mA).

Like you, your radio does not function well when it goes thirsty, so let's use an example that combines amps with what we've just learned about volts.

When using a mobile radio designed for use in a car, in the home, you can also use a rechargeable 12 V battery. However, HAMs find it more convenient to use devices called power supplies.

A power supply will draw standard 120 AC input from a wall outlet in your home or office and convert it to a stable output of 13.8 V. (Remember, 13.8 V is what your car's alternator produces when charging your battery.)

Power supplies are rated in amps. Obviously, the more amps a power supply can put out, the more power it can support. This is why the cost of a 120 VAC power supply will be commensurate with the number of amps it is rated to output.

Now, let's see the kind of power supply needed to run our example mobile radio as home base station unit.

Our example mobile radio has a transmit draw of 20 A. So, to properly provide continuous power from a 120 VAC source for this 12 V mobile radio, you'll want to use a 25 to 30 A power supply. However, this power supply can cost well over $150.00 so what about using a less expensive power supply such as an 8 A supply will cost about a third of a 30 A unit.

Will both the 8 A and 30 A be able to provide continuous power from a 120 VAC source for our example 12 V mobile radio? Only in standby and receive modes. When transmitting, here is where the 8 A power supply will fail the radio without an additional power source.

In order to to continuously power our example 12 V mobile radio with an inexpensive 8 A power supply, we'll need to use it in conjunction with a 12 V rechargeable battery. In this case, all three devices, the radio, power supply and battery are connected together. Here is the strategy.

When our example radio is in standby mode, that means it's turned on and ready to operate. This is the lowest power drain.

Our example radio draws 1.5 A in standby mode, but when it goes into receiving mode, the drain increases marginally. For example, the 1.5 A standby drain will increase to 2.0 A in receiving mode. In either case, an 8 A power supply can easily handle both the standby and receive mode drain from our example radio on its own.

However, the moment we key the microphone to transmit with our example radio, the power demand of the radio increases considerably. How much depends on the setting.

Most all amateur radios allow the operator to set the transmit power from low to high. Obviously, the drain for low power transmit will be less than than high power.

When evaluating radios for purchase, knowing the low and medium power drain is helpful, but the maximum high power drain is the specification you absolutely need to know!

In this case, our example mobile radio is configured for its maximum transmit power of 50 watts and drains 20 amps in this mode. Assuming we use it with an 8 A power supply where will the other 12 amps needed for the transmission come from? A battery.

Remember, the radio, power supply and battery are all connected together and when the power supply is on and the radio is in standby mode, the excess amps from the power supply are being used to charge the battery and maintain it in a charged status.

When we key our example radio to transmit, the first 8 amps of 20 are provided by the power supply and the other 12 amps come from the battery.

But this raises another question. For how long can you do this? That depends on the battery.

Rechargeable batteries such as those used in motorcycles, ATVs and so forth will be rated in terms of amp hours (Ah). So, let's assume you're using a small gel cell battery with this configuration, and it is rated for 6 Ah. In this case, you can transmit continuously for only about 15-30 minutes or so, depending on the battery and its condition.

If you want to gossip for hours on end, get a 30 amp power supply for your home base station. That is for now. After the global tribulation begins, you will find yourself short of gabbing power because the wonderful ease and stability of a 24/7 power grid will become a cherished memory.

In this environment, you'll be fighting for every amp you draw and will never forget it.

DC Power Management

Present day thinking is dominated by our easy 24/7 access to 120/220 AC service for our homes and businesses. During the global tribulation, there will be no grid to plug into for 24/7 AC power. For this reason, you must employ a DC power plan with all of your communication systems.

With this in mind, there are three power planning factors to keep in mind:

- **Establish Needs:** Anticipate your short term and long term needs for generating, storing and expending power in the most efficient ways possible.

- **Do the Doable:** Your radio power strategy must fit within the overall power plan for your community so that it can be both practical and affordable given your circumstances.

- **Power Priority:** Establish a power priority for each communication system in your community as it is virtually assured that at some point your community's ability to generate power will be affected by equipment failure or natural circumstances.

While the power planning factors discussed above are malleable, one radio planning factor is not. That is the use of DC power, and there are four basic types of DC power radios for survival applications:

- **Handheld Radios:** Small handheld radios transmit with 2 to 5 watts on average and can use a range of power sources, such as AA and AAA rechargeable and non-rechargeable batteries and 12 V automotive batteries.

- **Mobile Radios:** Mobile radios transmit with 25 to 50 watts on average and will be the mainstay of your radio inventory. These radios are designed for car installations and rely on 12 V DC automotive batteries.

- **QRP Radios:** QRP is a designation for low power. These small radios transmit with 1 to 10 watts on average and may use the same power sources as mobile radios. The big difference is efficiently. As a crude rule of thumb, a QRP radio can work the same bands but operate for a week on the amount of stored power a mobile radio will drain in a day or less.

- **Base Station Radios:** As a rule of thumb, base station radios such as those you'll keep permanently in a bunker will be an advanced mobile radio or what is called a compact radio. Compact radios, as they are slightly physically larger than mobile radios, offer a robust range of built-in features. Either way, advanced mobile and compact radios will serve as your 12 V DC base station radios.

Why not use full-sized base station radios? They typically require a 120/220 VAC power source, which makes them unsuitable for survival purposes for four reasons: power efficiency and flexibility, bug-out portability and cost.

You can use wind mills and solar cells to recharge your batteries, but for 120/220 VAC, you'll typically use a gas-powered inverter generator, or large batteries with a standalone DC to AC inverter. In any case, inverters come with added costs, complexities and inefficiencies.

Remember, AC is ideal for carrying power over long distances. On the other hand, DC is good for up a to a mile but within that mile, DC is a more efficient way to allocate your power output than AC. But is there enough distinction here to make a real difference?

While the added burdens of AC inverters are easily sustainable when the grid is up or during a short term emergency, the global tribulation will last for approximately a decade. During this time, you'll be fighting for every amp you draw. Keep it simple. Keep it efficient. Keep it local. Use DC.

Also, full size radios that require AC power are physically larger, heavier and more complicated than the four types of survival radios described above. This means, if you need to quickly bug out to a new location, moving a base station with full size radios will be more complicated and risky than bugging out with advanced mobile and compact size base station radios.

And finally the third reason to avoid AC power to recharge full-sized base station radios is cost. A full-sized AC powered base station radio will cost as much as two or three DC-powered advanced mobile and compact radios with the same transmit power. Why the added cost? It's because of the features.

A full-sized base station radio will have all the bells and whistles experienced HAMs use to smooth out more difficult contacts, especially with those overseas. That being said, during the global tribulation, are you really going to spend all your time talking with someone overseas? It's not likely because with your strategic budgeting strategy, less is more. Less bells and whistles equals more radios.

5

Strategic Budgeting Strategy

Previously, we discussed the need for a three-tier community communications strategy. Let's quickly review that as you will need to create a budget that addresses all three levels.

- **Level 1 – Emergency Radio System:** Amateur and citizen transceivers are used for day-to-day, general community operations, regional emergency communications and long distance (DX) communication via Morse code.

- **Level 2 – Field Radios and Repeaters:** Communications conducted beyond the community's outer perimeter in support of intelligence, security, reconnaissance, and scavenging teams, using multiple types of transceivers and repeaters.

- **Level 3 – Command and Control:** Amateur and citizen transceivers are used by command post operators.

When creating a strategy, the following three guidelines will help you to save money, time and frustration:

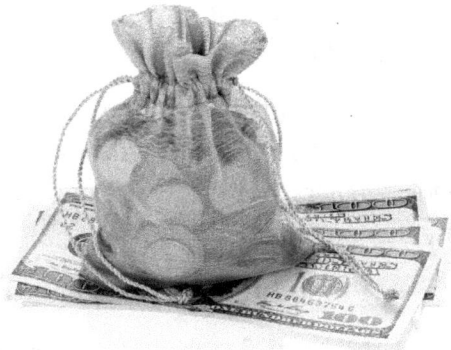

- **Bottom Up Priority:** The key thing to remember with strategic budgeting is that you build your budget from the bottom up, where level 1 requirements have the highest budget priority, level 2 the second highest and level 3 the third highest.

- **Be Practical:** With radios, you usually get what you pay for, and their range, operability, reliability and durability will be a direct measure of your purchasing considerations. Likewise, so will power efficiency and redundancy in depth.

- **Due Diligence:** Your final purchasing decisions will be a synthesis of your defined goals and the recommendations of your technical advisers. Here is where you can do research as a technical illiterate to evaluate recommended manufacturers and their products before you make a final buying decision with your team.

Further on in this book, we'll look at popular radios that are suitable for each level of your community communications strategy. However, the first thing we need to do is to view the budgeting considerations with an overview of the requirements for all three levels.

Level 1 – Emergency Radio System

If your emergency radio system consists of a few sealed barrels full of radios that must be retrieved and assembled in the midst of a crisis, what you actually have is a set up to fail system. This is why your emergency radio system must be used to handle daily community communications.

Radios that are power efficient, inexpensive and simple to use such as GMRS walkie-talkies are an ideal foundation for daily community communications and emergency needs. Other types of radios will be assigned for emergency needs as well.

How many of these radios do you buy? Let's assume you have a community of 100 people consisting of 25 families and that 20 percent of your community members have high priority needs for personal carry as opposed to shared access. Working from the bottom up, let's see how the numbers work out.

- **Personal Carry:** 20 radios equaling 20% of the community population.

- **Shared Access:** 50 radios, 2 per family.

- **Redundancy:** 210 radios total, based on 70 personal carry and shared family access radios with a 3X redundancy factor.

Also, keep in mind that you will need to power these radios, so do you calculate your power source based on 70 radios? No. You base it on 210 radios, which means a lot of batteries.

If you use rechargeable batteries, you, likewise need to factor in the power source, which could be solar cells, for example. Or, you could use both throw-away alkaline batteries plus rechargeable batteries. You are going to need a lot of batteries with most of them being rechargeable.

Note: Both rechargeable and non-rechargeable batteries must be periodically rotated, so be sure to check the expiration dates, especially with non-rechargeable, alkaline batteries for maximum shelf life. Also, never buy cheap no-name batteries. While they may claim a long shelf life, the big problem with no-name brands is leakage which can irreparably damage your components. For this reason, purchase reliable brands such as Energizer, Duracell & Rayovac, and be sure to factor in the added cost of a rotation plan for maximum shelf life.

As you can see, the cost of power for these radios will exceed the cost of the radios themselves. If you need to keep a modest budget, where is the wiggle room?

Order supplies in Case Lots. Standardization is not only essential for training and support, it also gives you volume purchasing power. Let's say you decide to give each family 1 radio instead of 2.

This reduces the number of radios needed to 135, which is typically sold in bubble packs of two or more. Assuming you buy 2 radio bubble packs, that comes out to 68 bubble packs.

For a modest budget, limiting each family to one GMRS radio could make a difference. However, these handheld radios will be your communication front line, and remember, emergency radios have the highest purchase priory.

So, what can you do to stretch your buying dollars enough to justify 210 radios so that each family can have two radios? Here are three things you can do.

- **Manufacturer or Distributor Case Lot:** While 68 bubble packs is a considerable order, 210 radios in 105 bubble packs can put you into a more favorable discount bracket. Plus, it will let you deal directly with the manufacturer or a major distributor as opposed to working with a deep discount retailer.

- **Be Mindful of the Power Source:** Let's assume you've narrowed your search for the ideal GMRS radio to standardize units for your community. Both semi-final choices have similar features but one uses AA batteries while the other uses AAA batteries. Here is where your choice of radio can make a big difference in your power costs.

  When you compare AA with AAA batteries, the difference between the two in terms of amps is considerable. An alkaline AA battery delivers 1800–2600 mAh (milliamp hours), whereas a smaller AAA battery delivers 860–1200 mAh. This makes AA batteries the value choice. For a small difference in cost, you get more than twice the amperage plus a considerable power source logistics savings. What's not to like?

- **OEM Offerings:** It is common practice for name brands to have their radios manufactured by OEM contractors. If you settle on a name brand radio but you're looking for a bit of savings, do some research to see if there is an OEM manufacturer for this name brand selling the same kind of radio. If so, the pricing could be more favorable. Also, an OEM manufacturer will be more willing to offer case lot pricing with drop-shipping.

Always remember: Each radio is a strategic purchase which means it must serve a specific mission role, and regardless of what radio you choose, its purchase cost will be just one portion of the overall budget. With small handheld radios, the long term costs will be influenced by the logistics necessary to enable them to serve their mission role.

Level 2 – Field Radios and Repeaters

One of the first things your security people will want to do when you're setting up your survival community is to establish defense perimeters. For example, they may insist that you create an inner defense perimeter by cutting back all trees and leveling the ground to a distance of 125 yards away from the community. This way, there is a clear line of fire that makes it hard for the nasties to sneak up on you. Also, because you've leveled the ground, the nasties will have to cover open ground without the benefit of defilade to protect them from your gunfire.

Likewise, your security team will ask for an outer defense perimeter manned by lookouts and sharpshooters stationed in observation posts. These observation posts will then support ground patrols in a wider area beyond your outer defense perimeter performing tasks such as scavenging sweeps, search and rescue, reconnaissance and sniping teams in natural terrain and disaster sites.

For these field operations, it is essential that you choose radios and supporting gear that are designed for field work, and you should spare no expense. You define the mission and then purchase the best radios for that mission. They will need to be rugged, simple to use and power efficient.

This does not mean that you need to spend money like a drunken sailor either, as there are affordable amateur radios built to military specifications that your security people will thank you for.

On the other hand, you'll also have other teams performing newsgathering tasks beyond the outer perimeter in other inhabited areas. Assuming your teams will be working in secure areas, here is where you can use more robust mobile radios designed for typical extremes. The trade off here will be less rugged designs that support more bands and features.

As with your level 1 radios, you want a 3X redundancy as well. Do not assume that these radios are more survivable than inexpensive consumer radios because the biggest threat you face is not someone dropping a radio. Rather, it is an electromagnetic pulse (EMP) from an extreme solar storm event that will fry most digital devices. Therefore, it may be prudent to store radios not in use within EMP-resistant storage.

Remember, no matter how well-designed a modern radio is, they're all made using digital components; while they may vary in terms of EMP survivability, no digital radio is totally immune to the power of the sun.

This is why you purchase radios with, at a minimum, 3X redundancy, and whenever possible, you should up the redundancy factor as well. After all, given that the global tribulation will last for approximately a decade, if a redundancy factor of three is good, six must be better.

Level 3 – Command and Control

In terms of features, base station radios will be more robust than those needed for levels 1 and 2. However, because you're using mobile and compact radios, the big difference will be in the supporting equipment such as amplifiers, antenna tuners, packet terminal node controllers (TNC) and so forth.

Because your level 3 purchases will be the third highest priority, a good deal of discretion is required because with this class of equipment, there are many things that you can do, and over time, you may want to try them.

For example, previously we talked about using earth-moon-earth (EME) technology to bounce your radio signals off the moon and back to a third of the earth. As much as you may want to purchase the necessary technology for this opportunity at the outset, the fact is, you will not need this technology until much further downstream, and most of it can be built with salvaged materials.

Here is also another thing to consider about all of these other advanced radio technologies. During the global tribulation, your scavenging teams will be going far beyond your outer perimeter to find the various things that you need to sustain your community. This is when you begin to look for equipment to support these advanced radio technologies.

At the outset, you need to spend money on mission-critical radio equipment for your base station, specifically, radio equipment that works to support your level I and level 2 mission roles. This means you will be working closer to home as your ability to hear within this range will be your immediate need.

Over time, you will begin reaching out with DX contacts across the country and overseas, but again, this is either going to be possible with technology your HAM operators have in hand, or technology you scavenge beyond the outer perimeter of your community.

As you can see, servicing the mission needs for all three levels of your community communication strategy will require a diverse range of radios and radio technology. As the leader of a survival community and the founder of your local Radio Free Earth network, you need to let your technical team sort through the technical issues while you'll need to be looking at the various choices from an equally-important, broad-brush level.

Due Diligence

Even trusted amateur and consumer brands can offer radios that disappoint buyers in some way. For this reason, when evaluating any radio, you need to perform your due diligence regardless of manufacturer and look for the following:

- **Country of Origin:** As a rule of thumb, the most reliable radios are made by American and Japanese firms. On the other hand, the big advantage of Chinese-made radios is cost. For example, you can purchase three or four Chinese amateur radios at the cost of a comparable amateur Japanese radio.

- **Trusted Reviews:** A few internet searches will reveal a vast number of review sites. However, the most authoritative source for amateur radios is the American Radio Relay League, ARRL, that publishes a product review column for their QST magazine. (QST is a radio code for "here is a broadcast message to all amateurs.")

- **Used Radios for Sale:** When you narrow your selection down to a handful of makes and models, start searching the Internet for used ones for sale. Popular, used radios will hold their value so it may be difficult to find used ones for sale. Conversely, unpopular or less reliable radios will be more widely available and at considerably reduced prices.

- **Large Special Offers:** Manufactures will occasionally run promotional offers on established popular models. These incentives can help someone on the edge of their budget to make a buying decision. However, if a less popular model is about to be discontinued or upgraded, the discount can seem quite attractive. Perhaps too attractive; keep that in mind.

- **Radio Message Boards:** When evaluating radio brands and models, take time to search user discussions about these products on Internet blog sites and message boards. Oftentimes, you can learn something about a particular radio that could prove to be a red flag in terms of your needs. Conversely, you could learn something about a radio that makes it even more attractive, given your needs.

A simple way to think of due diligence when it comes to buying radio technology for your community communication strategy is as follows. Rely on your technical team to help you decide on what to buy but rely on your own business acumen to help you evaluate who made it and who you're going to buy it from and on what terms.

As the leader of a survival community, your choice is simple. Do you want the best-marketed and least expensive radios, or do you want radios that work, and are you willing to pay for that proven reliability? The best way to avoid red herrings is to investigate merchandising claims.

Investigate Merchandising Claims

If you enter a search engine phrase for "best amateur radio," you'll see several listings for independent pages offering the top ten or fifteen radios within a category, such as HT or mobile. The list of radios will feature popular brands like Kenwood, Yaesu and ICOM but the majority of radios on the list will be inexpensive, throw-away Chinese brands.

- **Features Over Reliability:** The basis of these comparison lists is focused on competitive features as opposed to the reputation of the manufacturer and the proven reliability of the model.

- **Effective Internet Promotion:** The Chinese are extremely savvy in using the internet to penetrate markets with low-priced products. Inexpensive, throw-away

Chinese radio brands use the same marketing techniques that have been effective in penetrating other markets.

- **Prepper Targeting:** Popular name brands such as Kenwood, Yaesu and ICOM are awkward at best when marketing their radios into the prepper and survivalist niches. Rather than directly acknowledging the needs of this consumer demographic, they keep their focus on amateurs and for the most part, leave preppers and survivalists to sort things out for themselves. Here is where they have left the gate open for Chinese brands to market their throw-away radios to preppers and survivalists.

When sorting through internet sites for information, you should read the reviews and opinions by radio HAMs who will typically sign their messages with their own call signs.

Be careful when giving a review on a merchant's web site the benefit of the doubt. It is better to visit amateur HAM web sites, blogs and message boards and read the reviews of real people with real call signs.

Also, never assume internet reviews by anonymous individuals presenting themselves as preppers and amateurs are honest. Always verify the poster, and this includes assuring verified purchases on Amazon because it is easy for marketing groups to set up shell accounts for radio purchases so they can post flattering reviews.

If you are in doubt about a HAM review, you can use an FCC web site to look up the call sign. You enter the search string, "License Search-Federal Communications Commission," into any search engine, and this will guide you to the current URL for this call sign lookup service. This is a free service provided by the FCC, and you are not required to give any personal information to obtain the call sign listing. The FCC will give you the current license information on file for the call sign.

No matter who makes the advertised claim or how sincere a consumer product review may appear to be, do your due diligence, and do not give anyone the benefit of the doubt.

Above all else and if product reliability matters to you, the established reputation of the manufacturer is vital.

Work With Trusted Brands

While several different types of radios are featured in this book, the final mix will depend on your community situation, budget and environmental factors. However, as a rule of thumb, the fewer the brands you choose, the better. Therefore, you need to carefully evaluate each manufacturer. From a strategic vendor selection standpoint, the following manufacturer reputation criteria are helpful:

- **Commercial Lines:** The most demanding applications are commercial, military and paramilitary. While expensive commercial radios may be outside your budget, companies with a solid reputation in this area will have a culture of quality, and their organizational commitment to excellence will translate to less expensive models.

- **Mission Critical Reliability:** Do the manufacturers design their radios to high standards? All will say they do but you need to investigate how they design, and then rigorously test how their radios handle low and high temperatures, resistance to dust, water and vibration. Likewise, do they drop test their radios and check for other important factors such as spurious transmissions as defined by FCC-type acceptance rules?

- **Ergonomics:** The dominant design theme for a radio's user interface will center around its features or ease of use. Manufacturers that build radios with brand loyalty in mind will design them for ease of use with intuitive buttons and controls that are easy to access. They will have bright displays that are easy to read and visually well organized. Also, sound quality is absolutely critical. Does the radio sound good or does it sound good enough?

- **Brand Familiarity:** Consistency, in menu designs and the operation of different radios within the same brand, can be an advantage of brand familiarity. However, some name brand manufacturers will deviate from the menu and operations used in older models when adding a new radio to their product line. While a user interface deviation can make a new radio more competitive or desirable to consumers, it will come at the cost of brand familiarity and a higher front-end learning curve.

- **Software Programming:** Today's modern radios rely heavily on software-controlled menus. With the handheld, mobile and base station transceivers, a manufacturer's consistent goal of keeping a familiar user interface is highly desirable. In other words, if you are an experienced HAM and someone hands you a completely new radio, how soon must you open the manual to learn how to use a desired feature or mode?

Initially, these are basic criteria you can evaluate on your own, but in the long run your technical teams will fulfill a crucial role in finding the right mix of radios for your community. Therefore, as the leader of a survival community, your crucial parameters will be usability and reliability.

If your team opts for a complex mix of radios with an overall high learning curve, you will be the first to hear the bitter complaints. You should look for manufacturers that design new models using a familiar user interface. This will help reduce overall training demands and operational errors.

Whatever that is going to be, and regardless of the component choices, ordinary people will be the ones to make them work, and for this reason, user concerns must always be foremost in your mind.

Address User Concerns

For those of you who are old hands, you know that most of us who are amateur HAM radio operators tend to inherit our brand preferences from our Elmers who have helped ease us into this technological alphabet soup. Consequently, you'll need to learn the preferences of your technical team members as these preferences will likely influence their purchasing recommendations.

In the final analysis, what is the best radio brand or model for each of the three levels of your community communications strategy? That will be whichever will best serve your strategic communication mission requirements in a rugged and reliable manner as you will next learn.

The key to making this work is to keep the technology as simple as possible and to use it as often as possible to develop muscle memory.

The term muscle memory is synonymous with motor learning. This is a procedural process that programs specific motor tasks into memory through repetition. Military and paramilitary units rely on motor learning to prepare their warriors and responders to prevail under stress.

In a survival community, the same emphasis on motor learning is also necessary. For this reason, your emergency radios need to be used frequently and should be:

- Be simple to operate for men, women, and children;
- Be standardized for ease of use and minimal training;
- Be durable, reliable and effective.

If your community has a bottomless funding well, equipping it with commercial radios should be considered. If not, then finding a mix of more affordable consumer and amateur radios to satisfy your emergency radio requirements is the best approach. With this in mind, let's go shopping for Level 1 – Emergency System Radios.

Part 2 – Level 1
Emergency
System Radios

6

Handheld Transceivers

Low power, two-way handheld transceivers will serve your primary, day-to-day community communication missions inside your outer perimeter. They will also be used extensively by your field teams operating far beyond your outer perimeter.

Popular with both consumer amateur and commercial line-of-sight services, the various types fall under one of the two following designations:

- **Walkie-Talkie:** This term is used with consumer two-way analog handheld radios for the FRS/GMRS and CB services. As a rule of thumb, a walkie-talkie is an inexpensive, channelized transceiver.

- **Handy Talkie (HT):** This term is used with both analog and digital amateur and commercial two-way handheld radios for the VHF and UHF bands; HAMs prefer to use the acronym HT. As a rule of thumb, an HT refers to two different classes of handheld transceivers. These are digital and analog VHF/UHF amateur transceivers that require an FCC license for transmitting, and expensive, commercial digital and analog handheld transceivers that use special purpose bands such as the 800MHz "digital dividend" spectrum.

  With commercial HT transceivers, the institution or business that purchased the equipment will require some form of licensing although the actual users (such as employees and first responders) will be automatically covered by the institution or business license.

Of these various types, only two are suitable for survival over the extended course of a global tribulation. They are consumer (analog FRS/GMRS) walkie-talkies and amateur (analog VHF/UHF) Handy-Talkies. Both offer comparable voice quality and transmit power.

As to why the other types are not suitable for survival purposes, the big issue is not voice quality or transmit power. Rather, transmit mode and infrastructure dependence are the big issues.

 ● **Transmit Mode:** The authorized transmit power of a CB walkie-talkie is 4 watts but its real Achilles heel is that it works in AM mode, the least efficient mode for transmitting a signal. This inefficiency delivers range and voice quality that are noticeably subpar in comparison with consumer and amateur handheld transceivers which use the far more efficient FM mode.

 ● **Infrastructure Dependence:** As noted earlier in terms of survival, infrastructure dependence has become the Achilles heel of current state-of-the art commercial digital voice (DV) transceivers. While these transceivers are rugged and durable, they are also very expensive and depend on infrastructure such as a centralized trunking system or internet connectivity. During a global tribulation, these infrastructure resources will degrade and fail and thus, will cause similar results for infrastructure-dependent HT transceivers.

Now, we've narrowed the range to two options, consumer FRS/GMRS walkie-talkies and amateur VHF/UHF Handy-Talkies.

Consumer vs. Amateur Handhelds

GMRS UHF Consumer Handheld walkie-talkies and VHF/UHF Amateur Handheld Handy-Talkies should be compared on a value basis where the retail cost is balanced with the transceiver's degree of sophistication required.

 ● **Cost:** While prices vary with the brand and radio, a good rule of thumb is that for the price of one quality VHF/UHF amateur radio, you can purchase four or more well-made GMRS UHF consumer radios.

 ● **Frequencies:** Amateur HT transceivers offer a wide range of frequencies, whereas GMRS is a channelized service. With GMRS, there are only 22 channels, each with an assigned frequency. It is why GMRS transceivers are simple to use and thus, popular with outdoor enthusiasts.

 ● **UHF Band:** Both HAM amateur and GMRS radio handhelds use the UHF band for the best signal penetration through man-made structures. Amateur handy-talkies work in the 70cm UHF band, and GMRS walkie-talkies also work in the UHF band at about 65cm.

   UHF band radios are better suited for use in cities, survival community shelters and so forth because of their smaller wavelength. It is easier for their signals to pass through man-made structures, especially where reinforced concrete is concerned.

   On the other hand, single-band VHF and two-band VHF/UHF transceivers offer the 2 meter VHF wavelength which better suited for open terrain use.

- **Transmit Power:** VHF/UHF amateur and GMRS UHF consumer radios typically offer 5 watts of transmit power. With VHF/UHF amateur, this applies to all supported bands and frequencies. With GMRS, this depends on the channel used which further limits the number of full power channels.

- **FM Mode:** VHF/UHF amateur and GMRS UHF transceivers both work in FM phone mode. Consequently, they both offer a clear and bright listening experience.

- **Repeater Support:** Repeaters are used by HAMs to extend the range of their HT transceivers. All modern VHF/UHF amateur HT transceivers offer robust, flexible support for use with repeaters. Only a few GMRS transceivers offer repeater support, and setting up a GMRS repeater is difficult.

Despite the fact that both classes of radio transmit with 5 watts, the one thing the two do not have in common is range. While GMRS radios advertise a range of 35 miles, this is only under optimal conditions such as standing on a mountain top with a clear line of sight to another GMRS operator down in the valley below.

What is to be expected is that you will be communicating in an area with man-made structures such as a neighborhood range. In these less than optimal environments, a range of one to two miles with a GMRS walkie-talkie is more likely.

Conversely, expect an amateur handy-talkie offer to yield better working ranges with the same amount of transmit power because they offer more robust designs and better antennas.

However, the available range of a GMRS walkie-talkie is not a deal-breaker. This is because handheld radios will typically be used within the confined distances of a survival community where virtually most communications take place inside of a mile.

GMRS UHF Walkie-Talkies

At present, most General Mobile Radio Service (GMRS) radios offer both GMRS and Family Radio Service (FRS) services. The principal, technical differences between FRS and GMRS involve the transmit power which is available with each allocated channel, the ability to use repeaters and licensing.

- **FRS:** Depending on the channel, the maximum transmit power allowed is between 0.5 to 2 watts. The use of repeaters to extended transmit range is not available with FRS.

- **GMRS:** Depending on the channel, the maximum transmit power allowed is between 0.5 to 50 watts. The use of repeaters to extended transmit range is available with repeater-enabled radios.

The GMRS service does require a license. However, it is not necessary to demonstrate technical proficiency as with HAM amateur licensing. At present, new licenses can only be granted to a man or a woman as the head of household and will cover other household users. Businesses cannot obtain a new GMRS license.

When evaluating present day brands and models, you will see manufacturers are combining the FRS and GMRS services and selling them in multiple radio bubble packs. The result is that while it is perfectly legal to buy any FRS/GMRS radios without a license, only about 1% of all of these radios will actually be used with a license. Herein is a regulatory dilemma for the FCC.

Industry observers report a likelihood that the FCC will resolve this licensing issue by separating the FRS and GMRS services to prevent them from being combined by manufacturers. This will afford the FCC better control over the services and provide a significant increase in licensing revenues.

For the present time, whether you purchase a FRS/GMRS combined service or a GMRS single service radio makes no difference. Even if the FCC changes the allocations and rules for separating these services in the future, you'll still be able to purchase the combined service radios of today with full GMRS use well into the future.

Evaluating GMRS Walkie-Talkies

When evaluating FRS/GMRS walkies-talkie handheld radios, you'll find a diverse range of features and functionality. Let your technical team sort through these features but be sure to personally review the following criteria:

- **Power Source:** While the radios typically come with rechargeable batteries, you need to make sure they also work with AA non-rechargeable batteries and can be operated or charged with a 12 V source.

- **Ruggedness:** A handheld radio will spend most of its working life clipped to a belt where it will be exposed to the environment and abuse. For this reason, always look for durable FRS/GMRS walkie-talkie radios.

Most well-made radios can manage a splash of water (or blood) and a bit of blowing sand but the most important durability factor will be simple ruggedness. If you drop a radio off a shelf six feet up from the floor, will it survive the impact and keep working?

While manufactures make claims about ruggedness, these claims can be arbitrary. It is better to look for a liquid and dust industry rating as this is more reliable than claims. A design with a high rating will be inherently more rugged. These ratings will be discussed further on.

Whether your technical team feels it is useful or necessary now or in the future to setup a GMRS repeater to extend the range of your radios, your GMRS walkie-talkie radios must be repeater capable out of the box. There are no after market upgrades.

Therefore, while your technical team will be evaluating a wide range of features, the two features you must keep your focus on as a community leader are the power source and ruggedness.

If necessary, you can forgo repeater capability because levels 2 and 3 of your community communications strategy will include the use of mobile GMRS radios with a transmit power of 40 watts or more. This will be discussed further in the next chapter.

Motorola MS355R Talkabout

While there are many FRS/GMRS transceivers on the market, finding one that meets all of your requirements may be difficult. However, one walkie-talkie that is a good benchmark for conducting product evaluations is the Motorola MS355R Talkabout.

The MS355R is repeater capable and comes with a NiMH rechargeable battery that can go 9 hours on a single charge. Or you can use 3 AA batteries for up to 23 hours. One of its most impressive features is that it has a IP67 rating. It is not only waterproof, it floats as well.

When evaluating radios for water and dust, you'll typically see one of two industry standard codes, IP or JIP.

The IP Code (International Protection Rating) is commonly used with business radios to rate their resistance to both liquids and solids, and the Japan Industrial Standards (JIS) scale is commonly used with consumer two-way radios.

The IP Code uses two digits to classify dust and water, and the IP67 code rating for the Motorola MS355R Talkabout translates to:

- **Dust Rating 6:** No ingress of dust; complete protection against contact. This is the highest IP dust rating code.

- **Liquids Rating 7:** Ingress of water in harmful quantity shall not be possible when the enclosure is immersed in water under defined conditions of pressure and time (up to 1 m of submersion.) This is the second highest IP liquids rating code.

A detailed list of the IP and JIP codes is available within Appendix C – IP and JIS Specifications for Water/Dust Resistance.

The important thing to remember about GMRS walkie-talkies is that they offer the same 5 watt transmit capability on UHF frequencies as amateur HAM handy-talkies, albeit with far fewer frequencies and in many cases, less range.

What you may find practical is to use both. For simple tasks inside your inner perimeter, GMRS walkie-talkies could likely do the job. For teams operating beyond your outer perimeter, amateur HT handy-talkies will offer more suitable functionality.

VHF/UHF Handy-Talkie (HT)

When evaluating amateur HAM handy-talkie (HT) transceivers, it is important to remember that you do not need a license to listen. You only need to a license to transmit. Also, no matter what type of amateur handy-talkie you purchase, all available amateur bands used by HT only require a Technician class license.

With this in mind, you will see four different amateur HT offerings based on the number of transmit bands available:

- **FM HT (Single Band):** This is typically a 2 Meter (144 MHz) VHF radio.

- **VHF/UHF HT Dual Band (aka Dual Bander):** This transceiver will transmit on both the 2 Meters (144 MHz) and 70 cm (420 MHz) bands. These are the two most useful bands and the most popular type of handy-talkies used by amateurs.

- **VHF/UHF HT Tri-Band (aka Tri-Bander):** These radios are like a dual-bander with the addition of a third band which can be 1.25 Meters (222 MHz) or 6 Meters (50 MHz.)

- **VHF/UHF HT Quad-Band (aka Quad-Bander):** These transceivers have one of everything - 2 Meters (144 MHz), 1.25 Meters (222 MHz) 70 cm (420 MHz) and 6 Meters (50 MHz).

For survival, a popular dual band UHF/VHF HT is the best choice. Tri-band and quad-band HT transceivers are considerably more expensive because they are often duty-rated as submersible (as in you need to protect your investment.)

Is having the ability of tri-band and quad-band HT transceivers to send and receive on the 1.25 Meter (222 MHz) band or 6 Meter band worth the extra cost? No, because these are not highly active bands and because tri-band and quad-band HT transceivers will transmit on these bands with very low power levels. Therefore, if you want to work the 1.25 M and 6 M bands, more powerful mobile transceivers are recommended.

HT Radios for Survival Applications

Of the four different band allocations available with the Technician class license, the first most popular one in use is the VHF 2 Meter band followed by a close second, the UHF 70 cm band. The other two bands, 1.25 Meters (222 MHz) and 6 Meters (50 MHz), lag far behind in popular use. Therefore, you need to focus on rugged, time-tested dual-bander HT radios.

As the leader of a survival community and the founder of a local Radio Free Earth network, here is where you may find yourself at cross purposes with some members of your technical team who are looking for features and price value, and so, may be willing to consider a wider range of offerings. Obviously, here is where you will likely have your first conversation about Chinese radios.

Chinese manufacturers are heavily targeting the prepper market in America. You'll often see a picture of a prepper in camouflage gear holding an AR-15 rifle with a Chinese radio clipped to a shoulder harness. It may look convincing but the real truth comes out when you ask a HAM radio operator about Chinese radios.

If you ask an amateur HAM operator about Chinese radios, do not be surprised to hear them chirp, "Oh, yes, although, I do have a whole drawer full of dead ones. But hey, that doesn't matter because they're so cheap you can just throw them away when they fail. What's not to like?"

This may be a sensible strategy in a time when we can just order inexpensive replacements over the internet but during the global tribulation, do you really want to equip loved ones with unreliable radios because you can buy them for much less than a reliable radio?

If you're itching to save money with throw-away radios, then ask yourself, what is more desirable: being a clever but dead, band-for-the-buck, bubble pack consumer or a live survivor who did it right the first time (DIRFT)? Your answer depends on buying reliable HT radios.

Here is where quality radios that offer the latest and greatest in features can be less desirable than older radios with a solid legacy of reliability. To demonstrate a legacy of reliability, we'll use the Yaesu FT-60R dual-bander, the most recommended HT by seasoned HAMs for use by those new to amateur radio.

The Yaesu FT-60R

First introduced in 2004, the Yaesu FT-60R has been a long-time favorite of many HAMs who are certified members of the Amateur Radio Emergency Services (ARES), operated by the ARRL in association with Homeland Security. It obviously lacks some of the newer digital display and mode features, yet it is remains the most recommended, affordable HT for new HAMs.

- **Simple to Operate:** More modern HT radios place a heavier emphasis on software which means more programming needs to be done. This older radio gives operators more direct access to commonly used features with simple command sequences. One example is squelch control.

Squelch is used to suppress annoying channel noise sound when the radio is not receiving a significantly strong transmission signal. With many modern HT radios, this feature is now controlled by software where a command programming sequence is used to choose between low, medium and high squelch. That is good for now but how about during the during the global tribulation?

Under extreme survival circumstances, the fewer buttons you have to push to get something to work, the better, and the ability to control squelch will be even more critical than it is today. Ergo, in such circumstances, a simple thing like a squelch knob is far more preferable than having to push a sequence of buttons under pressure.

Here is where the Yaesu FT-60R offers the simple virtue of three knobs on the top of the radio. The first is an on/off and volume control knob. Then next to that are two other knobs, one sitting atop the other. The tall knob is for frequency tuning and the shorter and wider knob beneath it is for squelch control. All three knobs can be managed in the dark without having to look at a programming display.

- **Wide Band Reception:** Most every modern HAM radio offers a scan feature to help you automatically scan frequencies to find active contacts, and wide-band reception is a very desirable scanning feature.

  However, remember these two things about scanning. A dedicated scanner will do it faster than a transceiver in scanner mode and for less money. Secondly, when you use a transceiver in scan mode, it ceases to be a transceiver until you switch modes again.

- **Time-Tested Durability:** The Yaesu FT-60R was first introduced in 2004 with a rugged die-cast, water-resistant case construction. However, it is closer to the end of its market life cycle than the beginning, so if you're looking for the latest bells and whistles, you'll want to look at newer radios. Nonetheless, the FT-60R is a tough radio with time-tested durability, and more than one HAM has expressed amazement at having dropped one on a concrete floor without breaking it.

- **Price Point Retention:** As the leader of a survival community, price point retention is one of the things you must never overlook regardless of the brand or radio because this information is very revealing and easy to find.

  Even the most prestigious manufacturers have their own ugly ducklings and finding them is easy. Look for big sales discounts and loads of used ones at bargain prices on the used market. They are dead giveaways for ugly ducklings headed for the stew pot.

  While the retail price of the FT-60R has dropped slightly in response to competition from newer models with more advanced technology, the FT-60R continues to hold a good retail price since its introduction in 2004. When you find one on the used market, you'll see that it holds a respectable used price. Both are the signs of a time-tested and respected design.

- **After Market Accessories:** The other thing you must not overlook as the leader of a survival community is the availability of after market accessories for any radio you're evaluating. When a proven design has been on the market for several years, third party manufacturers will begin offering a wide range of after market accessories designed expressly for it.

  Such is the case with the FT-60R. Like price point retention, the availability of after market accessories is the sign of time-tested design. Plus, they also give you the ability to serve many different survival missions with the same basic radio.

This brings us down to the crux of it all. Is the Yaesu FT-60R the right radio for your survival community? This is up to you because your buying decision must be the outcome of the internal evaluations you'll conduct with your technical team. Let your team look for value. You look for consensus.

For this reason, the Yaesu FT-60R offers you as a leader of a survival community, a respectable benchmark for your evaluations, one that will help you keep your evaluations from drifting off into the weeds with sexy bells and whistles at the expense of time-tested reliability.

While the cost of an amateur Yaesu FT-60R HT may be several times that of a consumer bubble pack radio like Motorola MS355R Talkabout walkie-talkie, there are other reliable amateur HT models that offer more features than the Yaesu FT-60R HT at several times the cost.

Again, the Goldilocks principle applies. Not too much. Not too little. Just what you need. This is not only true for the selection of your handheld transceivers but is true for all of your other radios, as well.

In the next chapter, you'll see what is available with more powerful Citizens Band mobile transceivers, which, as a rule, cost about as much as a Yaesu FT-60R HT. However, they can transmit and receive messages with a range measured not in the tens of miles but in hundreds of miles, if not further.

7

Citizens Band (CB) Transceivers

Handheld transceivers will be necessary for routine communications between the members of your community. However, circumstances may force some or all of your community to relocate on short notice. In such cases, a rapid egress with a convoy of vehicles may be necessary, and handheld transceivers will be helpful in this role. If the route of the convoy follows established roadways, there will be other vehicles on the road as well.

During the global tribulation, we can expect to see road blocks on major thoroughfares, especially when routes cross a state line, in order to prevent floods of survivors from pouring into areas deemed essential to such uses as farming and ranching or to prevent irreparable damage. These roadblocks will not be established for voluntary public participation. Rather, you can expect those in power to impose the threat of deadly force, to arrest people, confiscate their vehicles and supplies, and if deemed necessary, shoot to kill.

For this reason, adding Citizens Band (CB) mobile transceivers to your level 1 – emergency radio equipment stores will serve you well. Relatively inexpensive and ubiquitous (found everywhere), they will serve vital roles in each of the three levels of your community communications strategy.

The reason that CB radios are valuable is that during the global tribulation, long-haul truck drivers, farmers and ranchers will become valuable sources of useful intelligence and friendly chatter about what is happening in the world beyond your line of sight or even around the next bend in the road. They also are ideal for command and control of a relocation convoy.

With that in mind, what about CB walkie-talkies? Are they useful enough to include in your community communications strategy? Only to a very small extent, if that. This is because the performance of the built-in telescoping antennas used by CB walkie-talkies is generally poor to fair. Hence, a popular expression shared among CB enthusiasts back in the 1970s was, "buy a $20 radio and an $80 antenna." The point being is that you want a mobile CB transceiver with a good antenna, not a CB walkie-talkie with a marginal antenna.

By some estimates, eighty to ninety percent of CB transceivers are of a basic type that only supports AM voice mode. This basic type, whether it be a mobile or walkie-talkie transceiver, is unsuitable for survival. The suitable type CB is called a Single Sideband Citizens Band (SSB CB) mobile transceiver.

CB and SSB CB

Look in the cab of an 18 wheeler and what you can expect to find is a basic AM mode only 40-channel CB mobile transceiver. However, when buying CBs for your community, you will need to break rank with most CB enthusiasts because as there are two types to consider, the basic 40-channel CB radios that most operators favor and 120-channel SSB CB radios.

The great thing about an SSB CB is that it is like getting two very different radios in one convenient box. On one side, there is the basic AM voice mode version that is used for short range communication by 80 to 90 percent of all CB users. On the other, there is the SSB CB version which can transmit internationally when properly configured.

Keeping this in mind, let's take a much closer look at CB technology beginning with a basic 40-channel CB. These transceivers only work in AM phone mode with 4 watts of transmit power, and all 40 channels are assigned to frequencies in the 11 meter band from 26.965 MHz to 27.405 MHz.

For long-haul truckers, basic 40-channel CB transceivers work well for their needs. However, the range is limited to a few miles at most. Also, when compared with FM mode, AM phone mode is not as clear and is much more susceptible to interference. However, if you're using a 40-channel CB with SSB, you're operating in a whole new world of possibilities.

SSB CB Modes

Single side band (SSB) comes in two modes, Upper Side Band (USB) and Lower Side Band (LSB). Ergo, while a basic CB has 40 channels, the SSB modes increases the number of available channel variants from 40 (AM only) to 120 (AM and SSB) as follows:

- Channels 1-40 AM phone mode;
- Channels 1-40 Upper Side Band;
- Channels 1-40 Lower Side Band.

Another benefit is transmit power. With AM mode, the FCC only allows 4 watts, whereas, a CB in SSB mode can transmit with 12 watts. Is 12 watts enough to make a difference? Absolutely.

With CB in basic 40-channel AM mode, you'll be lucky to make a contact across town without someone talking over you. However, once you begin using one of the SSB modes, you'll be working SSB, the most efficient transmit mode available on any radio frequency or band.

While basic CB radios are very limited in range in AM mode, once you switch them to an SSB mode, you will now have the DX propagation capability of bands in the HF region. For this reason, SSB CB radios offer the least expensive and easiest way to obtain the DX range capabilities of the HF region. As a result, many SSB CB operators in America report long distance (DX) international contacts in Canada, Mexico and Europe.

While the FCC does not allow international SSB CB DX contacts, there is no significant enforcement effort. Hence, SSB CB operators routinely make international DX contacts. Yes, they get away with it but only as long as they're not doing something outrageously illegal that will result in a loud knock on the door, such as using illegal transmit power levels.

The FCC has prosecuted more than one CB enthusiast who thought that the world needed a 1,200 watt CB station and then learned the hard way that prison food is truly awful.

Once the global tribulation is under way, the FCC's enforcement power will fade. That is when adding a linear amplifier to your base station SSB CB will be a good way to reach a broad audience of survivors.

In the meantime, you need to evaluate the newest SSB CB radios with an eye on a few helpful features.

Desirable SSB CB Features

SSB CB mobile transceivers were first introduced in the 1970s, and there are many inexpensive, used mobile transceivers on the market. But whether you are looking to buy an older, used SSB CB or one of the newest models available, look for the following features.

- Ease of use (Controls are ergonomically placed and easy to spot.)
- Sensitive receiver with strong frequency stability in all three modes, AM/USB/LSB
- Robust scan function in all three modes, AM/USB/LSB
- Bright and large digital display with selective day/night color schemes
- Built-in Signal and Standing Wave Ratio (SWR) meters
- Microphone gain (This can be thought of as a volume control for your microphone.)
- Public Address (PA) with a speaker output
- Battery check and internal diagnostics

There are many other desirable features such as weather channels. However, the huge benefits of SSB CB transceivers are their propagation range and comparability. Older CB and SSB CB radios may have only the lower 23 channels vs the 40 channels found in radios from the late '70s to current models. Several excellent SSB CB radios are now coming into the market, and one model, the Uniden Bearcat 980 SSB, is a strong contender that will serve as a reliable reference model for your evaluations.

Uniden Bearcat 980 SSB

While SSB CB radios are typically manufactured in places like China and Vietnam, they are designed in the USA with American operators in mind. Four such radios include:

- Cobra 29 LX CB Radio;
- Galaxy DX-959;
- President McKinley;
- Uniden Bearcat 980 SSB.

All of these radios are solid performers and offer great features for long-haul truckers. However, the Uniden Bearcat 980 SSB is the benchmark choice within this book for two important reasons. It has an excellent scan feature, and it does a good job of maintaining SSB frequency stability.

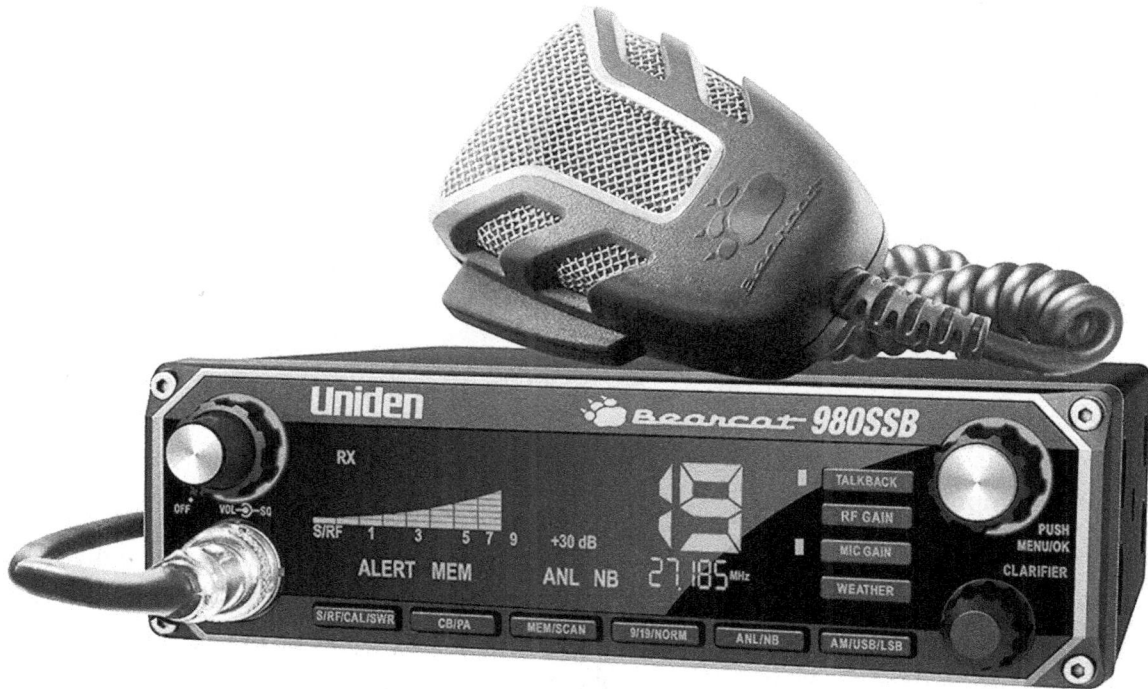

Regardless of the SSB CB radio you eventually choose, these requirements need to be at the top of the list. Other things to look for are ergonomics and built-in meters.

The simple ergonomics of the Bearcat 980 SSB make it easy to use for novices. The controls are well-laid out, the digital display is large, and it can be configured with different color schemes. It also offers a dimmer switch for day and night use.

Multiple built-in meters are always a useful feature, and the Bearcat 980 SSB has a large digital multi-function S/RF/CAL/SWR meter. The S and RF meters show transmit power and receive signal strength, and the CAL/SWR meters are for antenna tuning.

The CAL meter is for calibration and the Standing Wave Ratio (SWR) meter shows if your antenna is reflecting an excessive amount of RF (Radio Frequency) power back into your transmitter or not. As rule of thumb, you want want to keep the SWR under a value of 2.0 and within a target range of 1.1 to 1.5.

If your radio or antenna is not properly grounded, the SWR may be too high, and the result will be a loss of performance with a significant risk of damage to your radio. Never forget, antenna tuning will always play a critical role in working with any HF radio.

Now for that one desirable feature that gives the Bearcat 980 SSB its useful edge, the Uniden BC906W Wireless Microphone.

CB and SSB CB Microphones

One of the least popular aspects of most CB radios is the factory supplied CB microphone; this is especially true with newcomers after they hear what are known as "big audio" stations on the air. Consequently, they are quick to ask, "How do I make my CB sound louder on the air?" and "How do I get my radio to do more watts?"

What they learn is that if they want to be heard like a "big audio" station, there are three things they can do. Use a more expensive antenna, have a trained technician professionally tune their CB radio and buy an aftermarket powered noise canceling CB microphone such as the DM-452 Power Echo Microphone or the Astatic Road Devil.

Of the three, the one option most newcomers opt for is an after-market powered microphone. This is because using a powered microphone with your CB will help increase modulation through audio compression techniques.

As modulation increases, the radio's average power output increases; hence, your voice will sound louder for those listening to your transmission. Here is where the microphone connection becomes important. The factory microphone that comes with the Bearcat 980 SSB uses a 6-pin connector, whereas, popular after-market microphones use a standard 4-pin connector.

To their credit, Uniden supplies a 6-pin to 4-pin microphone adapter with the Bearcat 980 SSB, so evaluating alternative microphones for stock microphone replacement needs to be incorporated into your technical reviews.

However, there is one alternative microphone made by Uniden that enhances the survival communication mission usefulness of the Bearcat 980 SSB; it's the one you need to evaluate. It is the BC906W wireless microphone, and it works with three Bearcat models, the 680, the 880, and the 980 SSB.

While other CB manufacturers will no doubt eventually offer wireless microphones, as well, the BC906W is a first of its kind powered microphone. In fact, Uniden actually had to secure a waiver from the FCC to offer it. This is because the BC906W works in the 1920-1930 MHz UPCS band. Consequently, the BC906W operates in what is also known as the 1.9 GHz band, which resides within a wedge of allocated cellular PCS frequencies.

The BC906W comes in two parts, an adapter and a base unit that fits directly into the 6-pin connector with a battery-powered wireless microphone that has a talk range of 100 meters (328 feet) under ideal conditions.

Why use a wireless microphone with your CB or SSB CB? It would be for all kinds of reasons you cannot or would prefer not to imagine.

Chapter Review

When organizing a community communications strategy, several different types of FCC designated services will be employed. While most will be employed at one of the three levels of the strategy, the one service that will play a fundamental role in all three levels will be Citizens Band.

- **Recommended CB Type:** SSB CB mobile transceivers are the only Citizens Band service type that is suitable at all three levels of your community communications strategy for multiple survival roles. Basic 40-channel CB mobile and walkie-talkie transceivers are not recommended for new acquisition budgeting at any level. However, if you find them in garage sales at rock bottom bargain prices and they work, they may be useful down the line.

- **CB SSB Range:** Previously, we discussed three radio frequency regions, each with multiple bands and sub-bands. They are HF, VHF and UHF. Where VHF and UHF frequencies are limited to line-of-sight, bands in the HF region, such as the 11 meter CB band, have potential long distance (DX) propagation ranges of over 1,500 miles when transmitting in an SSB mode.

- **Existing User Base:** CB radios are still popular with long-haul truck drivers, farmers and ranchers. This affords excellent up-to-date information about road conditions, weather and other issues of importance, should a relocation convoy become necessary.

- **Channelized Ease of Use:** CB, SSB CB and commercial transceivers share a common attribute, the use of switchable channels (each assigned to a specific frequency), as opposed to tuning frequencies as is done with more sophisticated amateur transceivers. The result is that channelized transceivers come with a very low learning curve.

- **Affordable:** CB and SSB CB designs lack many of the complicated features HAMs use with more powerful amateur transceivers, such as those used to improve the reception of weak or distorted signals. This fact, combined with low transmit power levels and channelized ease of use, makes CBs and SSB CBs far easier and less expensive for manufacturers to build in volume. Given that they are mass-consumer products, these savings are passed on to the buyer.

- **Wide Availability:** It could be rightfully argued that the most common type of two-way radio available is the basic 40-channel CB, though it is limited in range. Popular with long-haul truckers, hundreds of truck stops and stores across the nation regularly stock and sell CB accessories. Add to that, there are vast numbers of older but serviceable transceivers in the used marketplace.

- **International Compatibility:** Regardless of the make, model or age, any basic 40-channel CB will work with any other basic 40-channel CB. Likewise, any 120-channel SSB CB will work with any other 120-channel SSB CB.

The bottom line is that an SSB CB is the easiest way to access the power of the HF spectrum via the 11-meter band with an inexpensive, reliable and easy way to access the long distance (DX) contact possibilities.

Next up is another HF region service type that will play a significant role in providing emergency communications and in building a culture of radio communication in your community. It is called QRP 40m CW.

8

QRP CW Transceivers

The term QRP refers to an amateur radio Q code that asks the question, "Should I reduce power?" With the advent of modern electronics, QRP has also taken on a new meaning. That is, to transmit and receive Morse code messages via HF frequencies in the CW (continuous wave) mode with very low power, over the longest possible distances.

To put this in perspective, ask yourself a defining question. Can you transmit a message from Los Angeles to London with a dual-Band HT like the Yaesu FT-60R, discussed earlier, with 5 watts of power? Without access to infrastructure, the answer is no.

However, simple QRP CW transceivers are completely infrastructure independent. Costing half as much as a Yaesu FT-60R, a properly configured QRP CW transceiver can reach London and beyond with 5 watts. In fact, there are amateur QRP HAM operators who love the challenge of making overseas CW contacts with 2 watts of power or less. Imagine that!

While more advanced QRP transceivers offer modes in additional to CW, the most popular and reliable mode for QRP is good old simple CW because a low-power QRP CW transceiver with ten or less watts can transmit Morse code messages across vast distances of thousands of miles.

This raises an appropriate question. What is the point of making contacts with people overseas during the global tribulation? After all, you cannot help them, and they they cannot help you, so why bother? In a word, the answer is hope. One of the deepest of human needs, it is not as easy to measure as common technical concerns such as transmit power and frequency.

As intangible and difficult as it may be, creating hope is how you will help survival community members with no interest in two-way transceivers to suddenly gain a genuine interest. But could this be an interest sufficient to motivate them for the difficult task of learning Morse code? For this reason, the third type of transceiver to complete your Level 1 – Emergency Radio System design must focus on building a family-based culture of communication.

A Family-Based Approach

If humanity is to emerge from the tribulation as a unified and free species, we will need three things; self-sufficiency knowledge, hope for the future and knowing that we are not alone.

These three things will be the universal needs of every survivor on the face of the planet, and most will be the young, especially teenagers; they will bear a heavy burden when the power grids and communication grids fail.

This is when they will lose touch with the very technology of a connected world that presently defines their lives, as many of the communication advantages of the last century fall by the wayside.

They will struggle with this new reality which will be made worse by the fact that the HT and mobile transceivers employed with any community communications strategy will require some form of rationing that will inherently favor the adults in the community. Consequently, when a youth is working a two-way radio, it will likely be under the supervision of an adult.

This is why the QRP CW foundation technology for your Level 1 – Emergency Radio System strategy needs to provide the young members of a survival community with an independent and satisfying form of participation, and it needs to be seen by all as essential to the viability and survivability of the community as a whole.

Do not assume that this QRP CW strategy is a convenient way to placate teenagers. Rather, this will be an opportunity to create a family-based culture of communication within your community. Not only will the young be involved, so will their parents.

While the most widely-used transceivers will be handhelds, QRP CW transceivers will be the most widely-used mobile transceivers. As such, they will become the foundation for a family-based approach to survival community communications.

During the tribulation, Morse code will become the new Twitter, a skill the youth of your community can learn and use to transmit messages over the air, or with signal lamps or by tapping on a wall, if it comes to that. When you implement this part of your system, the youth of your community will come of age knowing they'll always be connected to each other, even if the community becomes scattered to the four winds.

Imagine this. An unforeseen, natural event forces your community to split and relocate into separate groups. In this case, whether your groups are required to move a distance of a few miles or a few hundred miles, the ability to use QRP CW will enable the youth of these different groups to communicate over mountain ranges that will help instill hope in them to endure the misery.

To continue this example, see in your mind a group of young teens bidding tearful farewells to one another. As they prepare to leave, they all face each other one last time and proudly shout with a determined smile, "Catch you on the 40." This won't be a futile gesture.

Each will have or share a QRP CW transceiver, and they'll all be proficient and well-experienced in sending and receiving Morse code messages via CW. So when they say, "Catch you on the 40," they truly know they will.

Maybe they will never live to embrace each other again with warm hugs but their love for one another will endure, and they will be able to express it to each other for the remainder of their lives, even if only with through the monotone dits and dahs of CW on the 40m.

Catch You on the 40

QRP CW transceivers need to be less expensive and simpler to operate than the more sophisticated transceivers of your system. For this reason, they should be limited to one mode, CW (Morse code) which is the most efficient and reliable communication mode available.

Given the international range of a QRP CW transceiver, will language be an issue? No, because Morse code is English-based. Regardless of whom you contact and their native language, the conversation will always be in English. The real issue, then, is bands and frequencies, and the recommended HF band for QRP CW transceivers is 40m for the following reasons:

- **QRP Popularity:** QRP CW transceivers are available for most HF bands but the two most popular bands are 40m and 20m.

- **Licensing:** The 20m band requires a general class HAM license or higher so new operators with a technician class license cannot use the 20m band. However, the FCC has allocated 40m CW-only sub-bands for those holding a technician class license.

  Of the three license classes, technician, general and extra, the technician class license test is the easiest to pass and once granted, it also includes full access to all of the frequencies allocated to amateurs on the bands in the VHF and UHF regions.

  There are no age restrictions on becoming a licensed amateur, and after passing the examination, successful applicants are assigned a permanent call sign by the FCC. The cost of the license is minimal, and owning a radio or proving proficiency in Morse code is not required.

- **Propagation:** As a rule of thumb, the 40m band is good for local daytime use, and it offers excellent nighttime propagation for DX communications.

It is interesting to note the use of the 40m band by the volunteer HAMs of the Military Auxiliary Radio System (MARS) system which is a United States Department of Defense – (DoD) sponsored program with a long history of providing worldwide auxiliary emergency communications during times of need over most of the world.

MARS was originally created as the Auxiliary Amateur Radio System (AARS) in November 1925. Following the Cuban Missile Crisis, the AARS program was changed to the present day MARS program on 17 August 1962. Today, over 3,000 dedicated and skilled amateur radio operators provide the backbone of the MARS program.

MARS volunteer operators work both allocated amateur frequencies as well as DoD reserved frequencies. To work the DoD frequencies, MARS operators must provide proof of a MARS license to radio manufacturers in order to modify stock transceivers for access to restricted MARS frequencies.

Of particular note is the DoD requires participating HAMs to own a modern frequency-stable HF radio and antenna suitable for operating in the 3-10 MHZ range, covering the 80, 60, 40 and 30 meter HF bands.

By designating the 40m band for use with your QRP CW emergency radio system, you are making it as easy as possible for the youth of your community to become licensed HAMs with a technician class license and the privilege to use a strong HF band.

Unlike other bands, where a general or amateur extra license is required, the FCC has allocated CW-only sub-bands in the 40m band for those holding a technician class license but there is a catch. While HAMs holding a general or extra class license can transmit with a maximum of 1500 watts, technician-class HAMs are restricted to just 200 watts. Is this a problem? No.

With few exceptions, 100 watts will be the highest transmit power for the transceivers in your community communication strategy which is far more than the typical transmit power of a QRP CW transceiver that varies from less than one to ten watts.

To get your youth involved, they'll hone their QRP CW Morse skills with special training on extremely short range antennas that are ideal for indoor use. This way, they can safely communicate with each other. Eventually, they'll be ready to work regional, national and international DX contacts using more suitable antennas on their own.

Eventually, they'll want to work other HF bands, and this is to be encouraged. However, doing so requires a higher class license and a radio with a higher sticker price. This is why your best all-around option is going to be a single-Band QRP CW transceiver of which there are two basic types: single-band, crystal controlled QRP CW and single-band tunable QRP CW.

Single-Band Crystal Controlled QRP CW

Commonly referred to as "rock radios" by HAMs, single-band crystal controlled QRP CW transceivers have a long and distinguished history.

Before the 1920s, broadcast radio stations controlled their transmit frequency with tuned circuits. Hence, interference between adjacent stations due to frequency drift was a common problem. In the 1920s and 1930s, quartz crystal oscillators were developed to fix the drift by giving transmitters a high level of fixed frequency stability.

Today, single-band transceivers with quartz crystal oscillators, "rock radios," are still popular with QRP enthusiasts in America, especially those with an interest in preparedness. Consequently, many QRP CW transceivers are built and sold by amateur radio groups and experimenters.

However, a recent development is the importation of cheap QRP pixie rock radios (as they are called) from Hong Kong into the USA. While these are useful for experimentation and fun, they are poor choices for survival missions. Do not waste time on them.

Rather, evaluate American-based manufacturers and special interest groups such as the 4 State QRP Group based in the states of Arkansas, Oklahoma, Missouri and Kansas.

4 State offers a 40m QRP CW kit transceiver called the Bayou Jumper Transceiver that offers a noteworthy example of a QRP rock radio that pays homage to the famous spy transceivers of WWII with very efficient power drain in both receive and transmit modes.

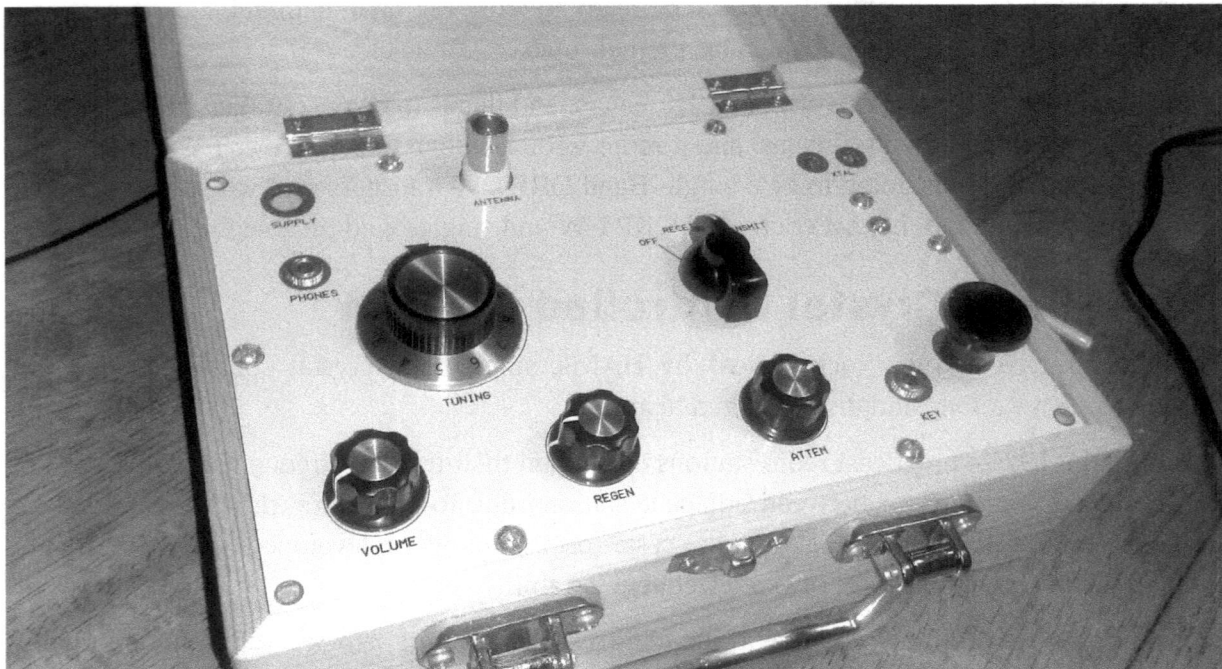

To illustrate the point, let's compare the Bayou Jumper with the Yaesu FT-60R Handy-talkie, described earlier, as both transceivers transmit a solid 5 watt signal.

4 State Bayou Jumper 40m Transceiver			
Input Power (VDC)	**Current Drain Amps (A) Milliamps (mA)**		**Transmit (Watts)**
VDC Input Range	**Receive Drain**	**Max Transmit Drain**	**Maximum Output**
12 V DC	10 mA	750 mA	5 Watt

Yaesu FT-60R HT Transceiver			
Input Power (VDC)	**Current Drain Amps (A) Milliamps (mA)**		**Transmit (Watts)**
VDC Input Range	**Receive Drain**	**Max Transmit Drain**	**Maximum Output**
7.2 V DC	125 mA	1.6 A	5 Watts

When you consider battery life as an issue, the power difference is substantial. The Bayou Jumper drain for receive mode is 1/12th that of the FT-60R HT and half as much power is needed to transmit a 5 watt signal. Yet, even with such a miserly power drain, the Bayou Jumper can get you across an ocean, whereas, you're lucky to get across town with a 5 watt HT.

A virtue of the Bayou Jumper is that it operates with only one quartz crystal oscillator at a time which makes it the most stable kind of HF radio to operate. Designed for the 40m band, the Bayou Jumper does not limit you to just one frequency. Rather, the frequency can be easily changed via the old style FT243 crystal plug on the front panel of the transceiver.

To change the frequency, remove the current crystal and insert a different crystal for the desired frequency. While it is getting more difficult to find older style FT243 crystals, there are ample crystal options with newer, smaller crystals than can be fitted via an FT243 adapter.

Two crystals are provided with the Bayou Jumper kit. They are 7.030 and 7.122 kHz, both of which are frequencies in the CW-only, 40 m sub-bands allocated by the FCC for HAMs with a technician class license.

Additional crystals for these sub-bands cost a few dollars each. In addition to the two crystals that come with the Bayou Jumper kit, you might want to purchase more crystals for the same sub-bands to include: 7.028, 7.040, 7.047 and 7.114 kHz. Now, let's get to the advantages and disadvantages of a crystal controlled transceiver.

Crystal Controlled Advantages

On the pro side, transceivers like the Bayou Jumper that use quartz crystal oscillators are simple to operate, highly stable and virtually free of drift.

Given that young teens will be using these CW-only transceivers without adult supervision, the immediate benefit of a crystal controlled transmitter is that they will only be able to send on specific frequencies using the crystals provided by the community. This is assuming they do not manage to scrounge up additional crystals on their own or drive you crazy by asking for more crystals.

The receiver on the Bayou Jumper is of regenerative design, is quite sensitive and is tuneable over most of the 40m band. Therefore, these crystal controlled transceivers are not limited to a single frequency which means your technical team can create a rock radio band plan for your survival community with a range of frequencies for the 40 m band.

Previously, we introduced six such frequencies, and we'll use them to illustrate one possible band plan to show how they could be assigned.

40 m Frequency	Description
7.028 kHz	Guard Channel – Continuously monitored for emergency traffic
7.030 kHz	Network Control – Dedicated channel for community traffic nets
7.040 kHz	Tactical 1 – General community use with DX antenna
7.047 kHz	Tactical 2 – General community use with DX antenna
7.114 kHz	Training 1 – Training channel used with short distance antenna.
7.122 kHz	Training 2 – Training channel used with short distance antenna.

With this band plan, you can teach young HAMs how traffic is organized to create effective communication environments using community-assigned frequencies and DX frequencies for long distance contacts. Also, note that the band plan shown above introduces two kinds of antennas for use with these transceivers: DX (long distance) and extremely short distance training antennas. Now, let's examine the con side.

Crystal Controlled Disadvantages

The Bayou Jumper is a hobbyist kit transceiver, and the operative word here is "kit." Kit transceivers (or kit radios) are, as the name implies, sold as hobbyist kits.

Rather than receiving a fully assembled radio that is ready for immediate use, kit transceivers come in bags of parts. In other words, not some but lots of assembly is required, and if you make a serious assembly error, your investment is basically wasted.

If you plan to build your kit, these transceivers are a good value for hobbyists with time to spare, the ability to solder and those on a sparing budget. That said, a discounted price is available when you purchase five or more kits, and various members of the 4 State QRP Group offer their services as QRP kit builders; they will assemble your transceivers on a fee basis.

If you decide to do it yourself, you will need to assemble 100 or so kit transceivers. For this reason, you'll appreciate the Bayou Jumper through-hole circuit board design which is much easier to

assemble than transceivers with surface mount devices. Also, you can purchase upgrades to the basic Bayou Jumper to increase its functionality.

However, between the cost of a basic kit, upgrades and kit builder services (if you want someone to build the transceivers for you), there are two serious concerns to consider, the first being time. One person sitting ona garage bench can only do so much – so quickly.

The second concern is cost. By the time you're finished, you may very well find that the production costs of a hand-made transceiver assembled from a kit builder service lack the benefit of lower mass production costs. In this case, transceivers that come assembled and tested from the factory are less expensive and more readily accessible.

Single-Band Tuneable QRP CW

A good example of a factory-made tuneable QRP transceiver is the MFJ 9040. This popular model has been on the market since 1991 and has a time-proven design; MFJ also makes other variants of the same radio for other bands, as well. What makes it tuneable is its variable frequency oscillator (VFO).

The first use of VFOs was for commercial and military applications dating back to late 1930. However, when the FCC made substantial changes in the band plan and allowed novice license holders to use VFO-equipped transceivers in 1972, the use of crystal controlled transceivers shrank, and now, they are used for specialty and experimental applications.

This is why virtually every transceiver on the market today, such as the MFJ 9040, uses a VFO of one kind or another. The four biggest differences between the Bayou Jumper and the MFJ 9040 are:

- **Cost:** What makes a transceiver such as the MFJ 9040 more expensive than a crystal controlled transceiver such as the Bayou Jumper is its VFO. That, plus the fact that the MFJ 9040 comes fully assembled from the factory, is why its retail price is roughly twice that of a basic Bayou Jumper. But you get what you pay for.

- **Construction:** Both transceivers offer rugged construction; however, you must purchase a separate case for the Bayou Jumper. The MFJ 9040 comes assembled with a case.

  The more expensive MFJ 9040 is built on G-10 double-sided, plated-through PC boards at the factory and is housed in a rugged all metal cabinet with brushed-aluminum front panel and vinyl-clad cover. Both are simple and durable.

- **Frequency Range:** A crystal controlled transceiver can only send on the frequency built into the crystals. The receiver is usually tuneable. To change the Transmit frequency, you change the crystal. For the MFJ 9040 (with the CW adapter), VFO allows those holding a technician license to send and receive Morse code message in the lower portion of the 40m band, from 7.000 MHz to 7.150 MHz.

- **Frequency Stability:** An important difference between crystal controlled transceivers and VFO transmitters and transceivers is drift. Crystal controlled transceivers are inherently frequency stable, whereas VFO transceivers require additional circuitry to maintain frequency stability.

One other notable difference between crystal controlled transceivers and VFO transceivers may be power drain.

QRP Power Drain

When it comes to power drain, a crystal controlled transceiver will hold a slight edge over a VFO transceiver. In many cases, there will be extra circuitry that consumes power. In normal circumstances with a working power grid, this is not a critical issue. However, during the global tribulation, you'll have to struggle for every amp of power.

With this in mind, let's compare the Bayou Jumper power drain with that of the MFJ 9040.

4 State Bayou Jumper Crystal Controlled Transceiver		
Input Power (VDC)	**Current Drain Amps (A)** **Milliamps (mA)**	**Transmit (Watts)**
VDC Input Range	**Receive Drain**	**Max Transmit Drain**	**Maximum Output**
12 V DC	10 mA	750 mA	5 Watt

MFJ-9040 VFO Transceiver		
Input Power (VDC)	**Current Drain Amps (A)** **Milliamps (mA)**	**Transmit (Watts)**
VDC Input Range	**Receive Drain**	**Max Transmit Drain**	**Maximum Output**
12 to 15 V DC	50 mA	1.2 A	5 Watt

As you can see here, both are 5 watt transceivers but the crystal controlled Bayou Jumper uses one fifth as much power in the receive mode as the MFJ 9040 and little over half as much power for transmit mode.

The power drain differences between the two transceivers is considerable but when you compare that with the other types of transceivers discussed within this book, one may wonder if this difference is a "tempest in a teacup," as they say. Or, is it the difference between life and death?

This is a tough call, indeed, and one that needs to be fully evaluated by you and your technical team.

Now that we've covered frequency stability and power drain, the third significant difference between the Bayou Jumper and the MFJ-9040 is key and keyer support.

Key and Keyer Support

In the early days of radio, the only Morse code key used was straight key, or what is also referred to as a telegraph key. These are still popular with HAMs today but there are also other keys (or keyers) with more advanced functionality:

Straight (Telegraph) Key
Iambic Twin Paddle Keyer
Bug (Semi-automatic key)
Single Paddle Key

- **Single Paddle Key:** The preferred keyer for Morse code competition, experienced CW HAMs use single paddle keys to win competitions with the highest scores in speed and accuracy.

- **Bug (Semi-automatic key):** Difficult to learn for many, the Bug is popular with a minority of HAMs.

- **Iambic Twin Paddle Keyer:** The most popular keyer in use today, it offers more speed than a straight key but the faster you key, the more you risk making mistakes.

With these more modern keyers, why should a straight key be the designated key for your level 1 emergency transceivers? There are five good reasons:

1. **Cost:** Straight keys are the least expensive keys to purchase as compared with the other types. This is because modern keys require electronic automation between the key and the transmitter. This can be added with an external or internal built-in keyer circuit which comes with additional power consumption.

2. **Simple Construction:** The least expensive straight keys are very simple in design and will use soft metal contacts. The most expensive will be sophisticated in design and will use gold plate contacts. And the cheapest key is two bare wires.

3. **Best Training Key:** Even HAMs who personally use advanced keyers agree that learning how to work Morse code with a straight key is the best way to begin.

4. **Ideal for Survival:** Survivors can easily fabricate a simple key with salvaged materials on hand, whereas, other keyers are too complex to be easily fabricated.

5. **Recognizable Fist:** With a straight key, the operator has full control over every aspect of the message transmission. The result is that straight key operators develop a "fist," as it is called. That is, the timing between dots, dashes, letters and words varies with each operator's fist and can be recognized.

Most widely used by CW HAMs today are the straight key and the iambic twin paddle keyer. While virtually all HAMs learn on a straight key, the number one reason why many learn how to use an iambic twin paddle keyer is a condition called "glass arm."

The term dates back to the earliest days of radio when operators worked for hours each day sending messages; the modern equivalents for "glass arm" are "repetitive stress injury" and "carpal tunnel syndrome." This painful condition is the principal reason for the development of single paddle, bugs and iambic twin paddle keyers which all have the same difference with the traditional straight key.

With a straight key, the hand and wrist are working up and down, whereas with more advanced keyers, the hand and wrist work side-to-side. For these reasons, all three levels of your community communication strategy need to address the availability of both straight key and iambic twin paddle keyers.

Determining Key and Keyer Support

The easiest way to see if a modern transceiver supports both straight key and iambic twin paddle keyers is to look on the back panel or front panel of the transceivers.

Modern transceivers are designed to support CW mode with a straight key and a monaural audio 3.5mm plug. Older transceivers will typically use a 1/4" monaural jack and may be powered with high voltage.

If the transceiver supports both a straight key and iambic twin paddle keyer, there may be a second stereo audio 3.5mm plug for the iambic twin paddle keyer in addition to the monaural audio 3.5mm plug for straight key support.

With this in mind, let's see how this applies to the two QRP transceivers described above.

● **4 State Bayou Jumper:** The stock version of this transceiver offers straight key support, and in the tradition of WWII era spy transceivers, it has a built-in simple key located in the front panel in addition to a monaural audio 3.5mm plug for an external straight key. Kudos to 4 State for a great straight key design.

To use an iambic twin paddle keyer with the Bayou Jumper, you must purchase an additional iambic keyer kit. This add-on kit, called the EZKeyer II, comes with a printed circuit board, microprocessor, memory and so forth. The assembled kit is very small and can easily be added to the transceiver with a second stereo audio 3.5mm plug for the iambic twin paddle keyer.

• **MFJ-9040:** This VFO CW transceiver comes stock from the factory with a monaural audio 3.5mm plug for straight key support and a stereo audio 3.5mm plug for an iambic twin paddle keyer with the necessary internal keyer electronics.

The factory standard iambic keyer support of the MFJ-9040 is functionally superior to the iambic keyer add-on kit for the Bayou Jumper.

As you can see by now, a stock transceiver is just the beginning which necessitates the need to share an insider secret about boats and transceivers.

Fancy boats make holes in the water that you fill with money. Two-way radios make holes in the air that you, likewise, must fill with money. This is just a heads up because there is another necessary hole to fill, your emergency QRP CW antenna.

QRP CW Antennas

In WWII, the two principal types of antennas used with HF CW spy transceivers were end fed and dipole wire antennas. These were simple to put up and highly effective for CW propagation. They were also quick to take down and easily stored.

That being said, those old WWII transceivers relied on vacuum tubes which were far less affected by improperly matched antennas than today's semiconductor-based transceivers. Therefore, matching the antenna to the transceiver is absolutely essential.

Here is where a single-band QRP CW transceiver with an end fed or dipole wire antenna constructed expressly for the band the transceiver supports can be easily erected in the field. Further, a dummy load antenna with a short piece of wire will do nicely for extremely short range training purposes.

Antenna meters and tuners are not required when an antenna is pre-tuned to the transceiver it serves. However, pairing a transceiver and an antenna tuned to dissimilar bands can cause severe damage to the transceiver. Never forget that!

Here is another point to keep in mind because it will help you keep the costs of your QRP CW transceivers to a minimum with a powerful do-it-yourself (DIY) approach.

DIY Antennas

Assembling transceiver kits with a myriad of small parts may not be practical. However, learning how to build simple pre-tuned wire antennas for your emergency transceivers is not only practical, it will be essential to your community communication strategy.

By working just one band, you can build and test a simple wire antenna for the transceiver. Assuming the two are properly paired, when you attach the antenna and power source to your QRP CW transceiver, it will be tuned and ready.

There are plenty of tuned 40m antennas available on the market today, and your team should obtain a few as reference models. However, building do it yourself (DIY) antennas for your emergency transceivers is a wonderful way to employ a family-based approach to this need.

Here is where you need to set the expectation with your technical team.

Today's antennas are made to endure specific environments for HAMs who typically erect them once and use them forever. These advanced antennas cost hundreds of dollars and will give you solid overseas contacts. However, the goal here is not to make it to the other side of an ocean. It is to make it to the other side of your county or state. Here is where a properly-constructed dipole or end feed antenna will do quite nicely.

Also, keep the environment in mind because during the global tribulation, hurricanes, typhoons, tornadoes, hyper storms, hyper velocity winds and blowing sand, grit and other particulates will tear even the most durable mast antennas into tiny metallic shreds.

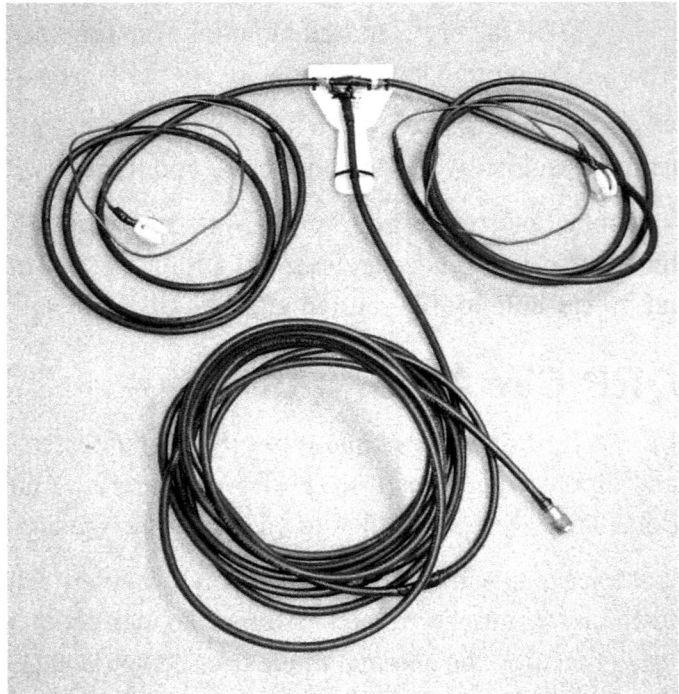

This is why you want to work with dipole and end feed antennas for your HF transceivers. They go up quickly and come down even quicker when danger approaches. Also, this strategy worked very nicely for WWII spies who typically ran their wire antennas through attic crawl spaces, as using a tall mast antenna was a great way to give their locations away.

Buying boxes of factory-assembled antennas for the global tribulation may seem practical but it is not. You'll go through them fast and then where will you be when they're all gone? This is why the right solution is to make sure your community families are self-reliant in building antennas on their own.

Family-Based Approach

Early on, you will need to instruct your technical team to acquire the necessary manufacturing supplies, tools and testing equipment to build your own DIY end feed and dipole wire antennas. At that point, the antenna guru on your technical team can design the antennas, construct test reference models for mass production and develop robust testing procedures.

Once this is done, you are ready to engage your families, particularly mothers because women are naturally better-suited to tasks that require deft assembly and attention to detail than men. If this comment offends the reader's political sensibilities, please ignore the advice.

For all others, here is where you begin creating a culture of communication within your community. Not only will moms build antennas, they'll fabricate the power cables and assemble complete emergency grab and go kits, as well. Consider this a true labor of love.

Once all the components necessary to work a level 1 emergency transceiver in the field have been gathered together, the final step is to organize all this into a grab and go kit.

Emergency Grab and Go Kit

Again, veering back to WWII, everything an allied spy needed was tucked away in a ubiquitous travel suitcase, including a transceiver, Morse code straight key, headphone, batteries and antenna. The same holds true for the global tribulation except for the use of a suitcase. Here is where a durable case such as a Pelican brand hard plastic case with foam padding is more suitable. These cases come with a configurable foam insert, starting near $40 and up, depending on size and configuration. Likewise, there are now a number of inexpensive knockoffs for much lower prices.

Once the manufacturing system is organized, each mom needs to work on each part of the complete emergency grab and go kit. She, likewise, needs to test it for reliability. In the process, she will learn how to make similar antennas for other bands such as 80m and 20m, for example.

Once this is working, the mothers of your community will build the very emergency grab and go kits their children will be taught how to use and to be responsible for. Yes, that's right. A family-based culture of communications is the ticket. Moms build them, and the kids operate them.

You now have the knowledge and tools you need as a survival community leader to work with your technical team in organizing the first foundation level of your community communications strategy. Next up is level two.

Part 3 – Level 2
Field Radios and Repeaters

9

Field Operation Transceivers

In terms of your three-tier community communications strategy, Level 2 – Field Radios and Repeaters is the most demanding because it will require a robust mix of transceivers (HT and mobile) and repeaters for extending your communication range over difficult terrain.

The scale of the actual number of components used and how they're organized will primarily be determined by the size and type of your survival community.

Small groups of survivors will organize using a "me-and-mine" approach to planning and preparation. Their plans are usually based on providing shelter and sustenance for at most ten to fifteen relatives and friends. Even with their use of sophisticated firearms, "me-and-mine" prepper groups will still be targeted more often by organized predators (such as gangs, cannibals, and militias) than larger well-organized survival communities.

Organized survival communities with populations of 100 or more members (with an ideal range being 150 to 200) will be more effective than the "me-and-mine" groups. These larger communities will operate both above and below ground and within well-defined and monitored inner and outer defense perimeters. The inner perimeter typically will have a clear- cut area that extends out from community structures roughly 125 yards or so. The outer perimeter which extends beyond that area will be determined by those in charge of the community's security.

The strategies developed in this book are based on the assumption that you are the leader of a well-organized survival community of no less than 100 members and that you will be organizing teams for various mission roles beyond your outer perimeter. Therefore, whether your field teams are concerned with security, intelligence, search and rescue, salvage, or news gathering, the communications gear you equip them with must be simple, effective, rugged, and as easy to use as possible.

Within the realm of commercial two-way radios this is a given for which you will give up a lot of money. It is more hit-and-miss with consumer and amateur grade two-way radios, and you're not without options.

As was discussed previously, analog type transceivers are recommended. Members of your community, however, may repeatedly inquire why you are not using next generation digital voice (DV) transceivers. This is a logical question and must be treated with respect because governmental agencies at all levels are making extensive use of DV and providing infrastructure support for trunking, encryption, and other services.

One explanation is the substantially higher costs of DV equipment; although that explanation can be easily sidetracked with a counterargument based on priority. Another could be that the resources and funds would have to come from other areas in the community's budget to finance the implementation of a DV strategy as opposed to a less-capable analog strategy.

Such questions are to be expected and not only will you need to address the financial drawbacks of DV for survival but the technical ones as well.

DV Pros and Cons

The radio industry is presently transitioning from analog to digital and the effort is a complicated work in progress. Hence the various offerings of the name brand manufacturers, Kenwood, ICOM and Yaesu, which fall into two basic groups. They are the last generation's analog transceivers such as the Yaesu FT-60R HT (discussed earlier) and new generational DV technology. This explains why few new analog models are coming to the market these days. Even so, the transition from analog to DV is not without some very serious teething pains.

The ongoing evolution of next generation DV transceivers is gaining speed, and with it comes the promise of a new and exciting evolution in amateur radio. Yet, this evolution has yet to progress far enough to address the long term survival demands of a global tribulation.

With this in mind the relevant pros and cons of DV transceivers are:

- **Voice Quality:** Digital Voice mode offers superb voice quality and resistance to radio interference. It delivers FM mode voice quality against a noiseless background.

- **Complex Information:** DV modulation when supported by infrastructure enables the exchange of more complex information than analog, through sophisticated DV services such as trunking and encryption.

- **Voice Over Internet Protocol (VoIP):** VoIP is a powerful feature of DV transceivers because it uses the internet as a trunk carrier. This feature eliminates the natural barriers of mountain ridgelines and summits. Here is how it works. You use your digital HT to connect via a local digital repeater; which then relays your transmission via the internet to another repeater running in the same digital network; where it is then transmitted to another digital transceiver anywhere in the world.

- **Lack of Standardization:** While there are attempts to standardize the modes and protocols for DV, the fact remains that in this early stage we see manufacturers promoting proprietary technologies. Consequently, emerging DV and VoIP radio technologies are locked up with self-serving maneuvers as manufacturers position themselves in this new marketplace. Many HAMs are concerned about becoming vested in the eventual losers before a winner is declared.

- **Design Complexity:** When a manufacturer creates a new radio for DV and VoIP radio technologies, the process is predictable. They essentially stuff a small computer into an analog transceiver for two reasons. Analog circuits are needed to transmit an RF signal. Secondly, when working with difficult contacts due to weak signal strength and/or interference, analog modes like FM and SSB may succeed where DV will fail.

- **Power Drain:** Analog ham transceivers are more stable than DV with regards to power consumption. This is a critical concern for survival, because how power is used is as important as how much of it is used.

  For example, analog modes work better for making difficult contacts and do with less power drain than DV modes. This could be a life or death difference in the midst of an emergency, when a transceiver is running solely off a battery source.

Radio manufacturers will eventually sort out these evolutionary DV problems and one or perhaps two universal standards will emerge triumphant. Until then, the number one downside of DV is infrastructure-dependence and the second biggest downside is power drain. Two of the proverbial 800-pound gorillas of survival communications.

In summary DV is about adding computers to HAM transceivers which increases their complexity, cost and power drain. Granted, all of these are acceptable downsides at present, but during the coming global tribulation, many of the technical benefits of DV will disappear into thin air – along with a good-sized chunk of your credibility as a leader.

So, when someone issues a challenge to your decision to use the latest generation of analog transceivers that begins with something like "well it seems logical to me that…." you have your answers and they'll be convincing. However, a solution will be even more comforting than a convincing answer.

Here you have a card up your sleeve. After you explain the pros and cons of DV, it's time to change the tune. Now you will talk about military standard amateur grade transceivers that are suitable for field operations beyond your outer perimeter. A universally respected standard more commonly known to HAMs as "MIL-STD (military standard)", "MIL-SPEC", or "Mil-Specs."

MIL-STD VHF Mobile Transceivers

When working beyond the outer perimeter, security teams will conduct patrols on foot and/or with off-road vehicles. In either case they will naturally prefer easy-to-operate single-band VHF transceivers. Both for their vehicles as well as for their own personal carry. Therefore, both types should conform to U.S. Department of Defense MIL-STD 810 C, D, E, F & G environmental standards for vibration and shock.

Why single-band (VHF) transceivers for open-terrain field operations? When working on open terrain, the band of choice for line of sight communications is 2 M VHF for both mobile and handheld transceivers. Here, there is very good news. Multiple manufacturers offer excellent MIL-STD rated models and for very reasonable prices.

Yet, you may wonder what the real difference is between a rugged amateur transceiver and one that has a MIL-STD rating? In a word, the difference is "extreme."

Amateur transceivers that are not MIL-STD rated are nonetheless designed to work in a difficult environment such as the cab of a truck, where it will be subjected to road vibrations, bumps and the typical temperature extremes of a motorized vehicle operating on paved or unpaved roads. Herein is the difference.

While a MIL-STD rated transceiver is designed for the same on-road operating environment, but it is also designed for off-road operations as well. In other words, an amateur MIL-STD rated transceiver is like a Timex watch. "It takes a lickin' but keeps on tickin'."

To help illustrate the difference between a rugged amateur mobile transceiver and a MIL-STD transceiver, we will compare two product offerings from Kenwood, the Kenwood TM-281A MIL-STD single-band mobile transceiver and the Kenwood TM-V71A dual-band mobile.

Amateur vs. MIL-STD Comparison

The Kenwood TM-281A is a rugged, MIL-STD VHF FM transceiver that puts out a respectable 65 watts of transmit power with choice of power settings.

To illustrate power drain, we'll compare the TM-281A with another Kenwood mobile, the TM-V71A dual-band.

Kenwood TM-281A – 144 MHz MIL-STD Mobile Transceiver			
Input Power (VDC)	**Current Drain Amps (A)** **Milliamps (mA)**		**Transmit (Watts)**
VDC Input Range	**Receive Drain**	**Max Transmit Drain**	**Maximum Output**
13.8 V DC ±15 %	< 1 A	14 A	65 Watts

Kenwood TM-V71A – 144/430 Mobile Transceiver			
Input Power (VDC)	**Current Drain Amps (A)** **Milliamps (mA)**		**Transmit (Watts)**
VDC Input Range	**Receive Drain**	**Max Transmit Drain**	**Maximum Output**
13.8 V DC ±15 %	1.2 A	13 A	50 Watts

As is shown in the table above, the power drain of the single-band TM-281A is comparable with the dual-band TM-V71A. However, because the TM-281A is a single-band model, the receive drain is 20% less. Furthermore, it only requires one additional amp for 65 watts of transmit power versus the 50 watts of the TM-V71A.

While the TM-V71A is a very rugged amateur transceiver and well suited to both on-road and light duty off-road mission roles, it does not offer the extra ruggedness of a MIL-STD transceiver such as the TM-281A.

However, the TM-V71A does offer a range of technical features that is substantially more than that of a TM-281A.

We will discuss the TM-V71A in depth later, but for now a comparison of dimensions and weight between these two Kenwood transceivers is important. In technical parlance, this is called "form factor."

The term form factor is used by the electronics industry to define the size, shape, weight, and other physical specifications of a component or product.

Mobile Form Factor

When used with an ATV or other off-road vehicle, mounting space will be at a premium. Also, if the TM-281A is carried in a backpack, dimensions and weight will also be important. Here is how the form factor for the TM-281A and the TM-V71A compare:

Radio	Width	Height	Depth	Weight

TM-281A	6.3" (160 mm)	1.69" (43 mm)	4.97" (126 mm)	2.49 lbs (1.13 kg)
TM-V71A	5.51" (140 mm)	1.69" (43 mm)	8.39" (213.1 mm)	3.3 lbs (1.5 kg)

In terms of width and height, the form factor of the two transceivers is comparable. But with depth comes a key difference. The TM-281A is 40% shorter and thereby easier to mount in a smaller off-road vehicle such as a 4x4 ATV.

Another important form factor difference is weight. The weight of the TM-281A is 25% less than the TM-V71A and while 0.82 lbs (0.37 kg) may not seem like a big difference with a vehicle-mount, it will if you're lugging a backpack for ten miles.

Another form factor advantage of the TM-281A is that the speaker is front mounted as opposed to being on the top or bottom of the radio. When operating in a noisy vehicle or open-cab environment, front-facing speakers make it easier for operators to hear messages loud and clear.

For off-road field deployments beyond your outer perimeter, a 2 M single-band high gain 5/8 wave whip antenna that mounts directly to the vehicle is recommended. This is because the propagation characteristic of this antenna type is more horizontal than vertical as opposed to other antenna types. Consequently, signals will travel further and voice quality will be better as a result of experiencing less interference.

For backpack use in the field, there are several different antenna options. This is where your antenna guru will need to match your requirements with the best antenna technology. In either case, mobile transceivers like the TM-281A with 65 watts of transmit power will enable line-of-sight communications with your field and base stations.

There will also be a need for communications between team members, and here you will want to use MIL-STD VHF handy-talkies (HT).

MIL-STD VHF HT Transceivers

When selecting a handy-talkie (HT) for use by teams operating in the rugged conditions of open terrain beyond your outer perimeter, the durability, form factor, and power are important criteria. With this in mind, let's look at the Kenwood TH-K20A which is a VHF single-band MIL-STD design that is easy to use and offers good reception and sound.

Durability in the field is a key advantage of the Kenwood TH-K20A. Like the TM-281A mobile discussed above, this HT also complies with U.S.

Department of Defense MIL-STD 810 C, D, E, F & G environmental standards for vibration and shock. It is also rated with an IP54 code for dust and water protection.

We previously reviewed the International Protection (IP) code for the Motorola MS355R Talkabout GMRS walkie-talkie, which is a respectable level IP67 for dust and water protection. The IP54 rating for the TH-K20A means that it runs a close second. The descriptions for the IP54 rating are shown below.

Kenwood TH-K20A – IP54 Dust and Water Protection	
IP Code	IP Description
5	Ingress of dust is not entirely prevented, but it must not enter in sufficient quantity to interfere with the satisfactory operation of the equipment; complete protection against contact
4	Water splashing against the enclosure from any direction shall have no harmful effect.

Given that the Motorola MS355R has an IP67 rating vs. the IP54 rating of the Kenwood TH-K20A does that mean it is more rugged? No, even though the Motorola MS355R has better dust and water protection, the vibration and shock resistance of the TH-K20A is the only model in this comparison that has a MIL-STD rating.

In other words, when members of a field team are on foot patrol and have to hit the dirt suddenly for any number of reasons, the shock that comes from suddenly hitting the ground is what the MIL-STD rated TH-K20A is designed to handle.

Kenwood is not the only manufacturer to offer MIL-STD amateur receivers. For this reason, you and your technical team should also evaluate the MIL-STD offerings of other brand name radio manufactures, and not only for their durability, but for the other criteria as well such as form factor.

HT Form Factor

When it comes to HT transceivers, the term form factor is not as clear cut as it is with mobile transceivers. Hence, the form factor on an HT can feature highly-subjective descriptions such as a brick, too tiny, and other such definitions.

To illustrate this issue, picture a HAM operator with a beer-barrel belly and a large HT radio clipped to his pants belt. The arrangement works fine as he strolls around town, but if he needs to run across a busy street, do not be surprised to see him holding onto his HT with a free hand while he dashes across the street. After crossing the street, he has to cinch up his belt and/or pull up his pants a little which is a minor but acceptable inconvenience.

With field teams, a minor inconvenience like this could become a deadly hindrance. This is why military and paramilitary personnel wear web gear with shoulder harnesses that comfortably distribute the weight they carry, especially during moments of high exertion.

To help frame this issue of form factor more clearly, let's compare the Kenwood TH-K20A single-band HT with the Yaesu FT-60R dual-band HT that we introduced in a previous chapter.

Radio	Width	Height	Depth	Weight
FT-60R	2.30" (58 mm)	4.3" (109 mm)	1.20" (30 mm)	13.05 oz (370 g)
TH-K20A	2.13" (54.1 mm)	4.4" (111.77 mm)	0.57" (14.48 mm)	7.4 oz (210 g)

As is shown in the table above, the two transceivers are a close match for width and height. However, with depth and weight, the form factor of the TH-K20A is clearly superior. It is 52% as thick and weighs approximately 46% less than the FT-60R.

To visualize this difference, imagine that you are taking hostile fire and running for cover. Which of these two transceivers would you want clipped to your shoulder harness?

That said; the many features and after market accessories available for the FT-60R make it hard to ignore even though it lacks the MIL-STD and IP durability ratings of the TH-K20A. For example the battery power source is where the FT-60R has the upper hand.

Power

As the leader of a survival community, you'll struggle for every amp of power for your shelters, electronics, and so forth. This is why one of the first power criteria mentioned was the differences between the Yaesu FT-60R and the Motorola MS355R where both can use AA alkaline batteries. On the other hand the Kenwood TH-K20A requires an exception to this standing rule because it uses AAA alkaline batteries instead of AA.

Out of the box, the TH-K20A comes with a rechargeable lithium-ion battery that takes about three hours to charge in the cradle and it will last up to 10 hours with a low power transmit setting. Kenwood also offers two other power options, the KVC-22 DC 12 V vehicular charger adapter, and an optional battery case that requires six alkaline AAA batteries.

**Kenwood KVC-22 DC 12 V
Vehicular Charger Adapter**

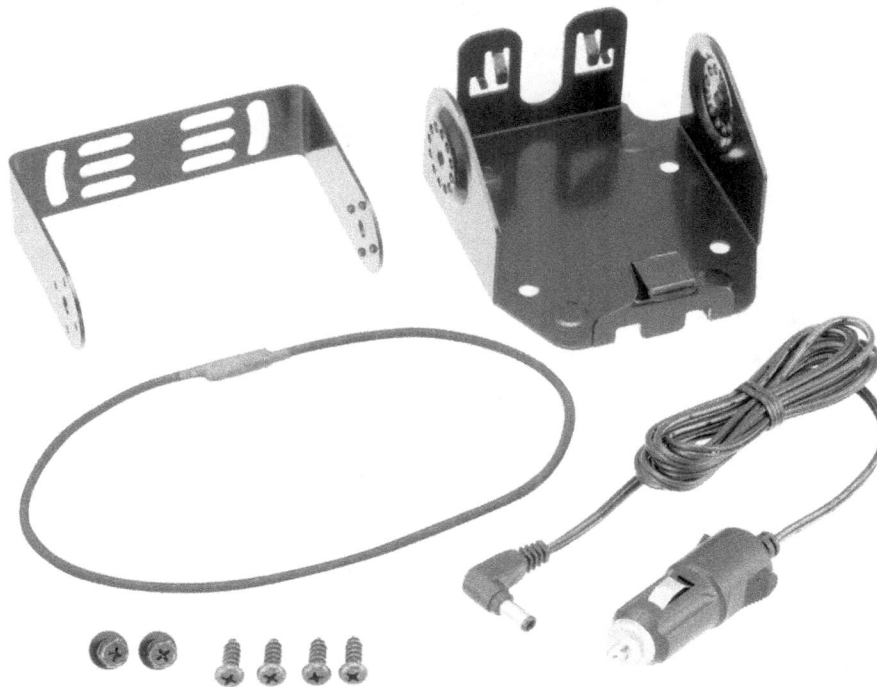

**Kenwood BT-16 Battery
Case a.k.a. "Clamshell"**

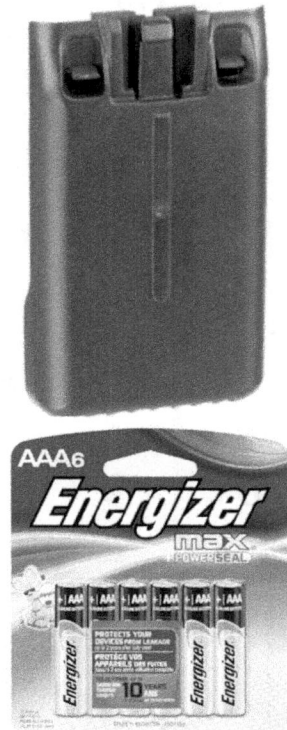

It is important to note, that with most handheld transceivers, the use of alkaline batteries requires lower transmit power. For example, although the TH-K20A is rated for 5.5 watts when using its rechargeable battery, it's transmit power with the alkaline battery case is 3.5 watts.

You may wonder why Kenwood opted for the AAA alkaline battery case for the TH-K20A. The answer is simple – form factor. As discussed above, the thickness of TH-K20A is 52% that of the FT-60R.

In fact, Kenwood made a company-wide decision to implement AAA alkaline battery cases for thinner and smaller form factors instead of performance with AA size batteries.

As the leader of a survival community, this presents a quandary for you and your technical team. Do you strictly enforce the AA rule which means going with a field operation HT with a larger form factor, more features and better power option? Or, do you make an exception to this power standard and favor of a lighter and thinner MIL-STD option? The answer is that you must consider the exception based on who uses it.

Field teams operating beyond your outer perimeter will not be concerned with more mundane issues such as plugging a leaky hydroponic tank. Rather, they'll be putting their lives on the line for the community. Therefore, if you respect their need to be confident in the gear they're taking into the field, explain the issues let and let them make the call. If they want a slimmer and lighter MIL-STD HT like the TH-K20A, then you must cheerfully make that exception.

But these two MIL-STD VHF transceivers are not the only things you'll need to acquire in support of your field teams that are working beyond a distant mountain ridge in the next valley. A 65 watt VHF transceiver designed for line-of-sight use, cannot get a signal over a mountain on its own – unless it is working with an appropriately located repeater.

10

Using Repeaters
for Distance

To work contacts over a mountain range, HAMs and commercial operators and associations use transceiver systems called repeaters. Principally used to extend the range of low-power HT transceivers, they re-broadcast transmitted signals at a higher wattage from a high location thereby allowing a much larger coverage area.

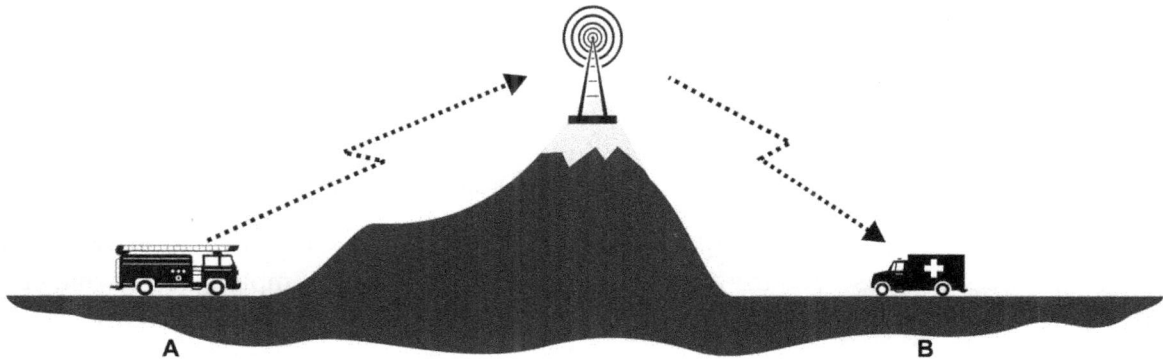

During the global tribulation, repeaters will be an essential tool for extending line-of-sight communications across one or more mountain ridges surrounding your survival community. However, expect many of the repeater systems currently maintained by commercial operators and associations to fail due to a lack of maintenance, fuel and any other combination of factors.

While restoring failed repeater systems back to working order may become an important community outreach mission for your technical teams, the more pressing goal is to create and maintain your own repeater system.

As repeaters are a unique type of communication technology and there are several variants. For this reason, a basic understanding of the technology is an imperative, beginning with the difference between simplex and duplex communications.

Simplex and Duplex Communications

With simplex communication, messages travel in a single direction. Also referred to as half-duplex, simplex signals flow in one direction at a time. Consequently, an operator can talk, or listen, but not simultaneously.

Therefore, direction is the key factor here and with simplex two-way conversations, the direction of signal flow is reversible via two modes:

- **Outbound:** In the outbound transmit mode a transceiver transmits the message via the assigned frequency, but cannot receive at the same time.

- **Inbound:** By changing a transceiver to the receive mode the direction is reversed on the same frequency, until the transceiver is switched back to the outbound transmit mode.

Always remember. Simplex (half-duplex) communications are always one-way and the transceiver is either in send or receive mode. It cannot send and receive simultaneously like a telephone, which is full-duplex. Transmit and receive occur on one frequency.

With duplex (full-duplex) communication, simultaneous send and receive is possible. This eliminates the need to manually reverse the message flow. With full-duplex, messages are simultaneously sent and received.

A good example of a duplex communication device is your cell phone or land line telephone. You and the other party can talk to each other and hear each other at the same time.

With this in mind, let's see how direction plays a key role with three different repeater technologies that are sure to useful during the coming global tribulation: simplex, duplex and crossband.

Simplex Repeaters

To create simplex repeaters, a transceiver is integrated with a simplex repeater controller. This is a versatile type of repeater as it can be quick and easy to set up in the field.

Three physical components are required for a simplex repeater. They are the controller, an adapter cable, and a compatible transceiver. When it is operational, the transceiver is set to the same frequency for both the receive and the transmit modes.

By integrating a simplex repeater controller with a low-powered handheld transceiver you can increase the overall signal range. However, use the same simplex repeater controller with a high-power mobile transceiver, and the range will be significantly more. Here is how it works.

Connect the Components: The transceiver is connected to the simplex repeater controller via an adapter cable that enables the controller to switch between receive and transmit modes. (Remember, never power up a transceiver without a properly configured antenna or dummy load attached.)

1. **Power and Set Frequency:** Power up the components and set the assigned frequency using the transceiver.

2. **Traffic Monitoring:** In receive mode, the transceiver continually receives (monitors) the assigned frequency for incoming audio messages from another operator using the assigned frequency.

3. **Receive Mode Operation:** Incoming messages are relayed via the adapter cable to the simplex repeater controller, which immediately begins storing the audio message in memory for re-broadcast.

4. **Send Mode Operation:** When the incoming message ends (or the controller's audio memory is full), the controller then commands the transceiver to switch from receive mode to transmit mode.

5. **Sending Stored Messages:** With the transceiver in transmit mode, the controller then replays the audio message stored in the controller's memory through the transceiver.

6. **Automatic Controller Reset:** After the stored message has been completely transmitted, the controller automatically clears its memory and commands the transceiver to switch back to receive mode. The repeater then waits for the next transmitted message which restarts the process.

When using a simplex repeater, remember that to send a message, first you will key your PTT (Push To Talk) microphone. Do not release the PTT switch until after you send the message. After sending your message, release the PTT switch and wait for the repeater to completely re-broadcast your message and reset to receive mode operation before sending another message.

The pros and cons of simplex repeaters for global tribulation survival applications are:

- **Simplex Pro:** With the correct simplex repeater controller and adapter cable, any transceiver can be configured in the field as a repeater without a lengthy setup process or the need for expensive electronic testing equipment to calibrate the repeater.

- **Simplex Con:** Simplex repeaters require the sending and receiving operators to wait for the replay to finish. This time lag needs to be anticipated. Also, additional equipment and power are required to operate the simplex repeater controller.

During the global tribulation, natural events such as an electromagnetic pulse (EMP) or normal wear and tear can cause full or partial transceiver failures. Consequently, your community may find itself with a small supply of multi-band transceivers that only work on one band or function.

In such cases, having a supply of simplex repeater controllers on-hand could provide a practical way of adapting partially inoperative transceivers for use as simplex repeaters. Be sure to discuss this strategy with your technical team, though they will likely want to focus more attention on setting up duplex repeaters for your community operations due to their ease of use, power, and range.

Duplex Repeaters

Simplex repeaters use only one frequency for both transmit and receive (input and output) - but not simultaneously. However, duplex repeaters use two separate frequencies for simultaneous input and output, which is why they are called duplex repeaters.

Many cities and towns across the country are served by duplex repeaters for amateur, commercial and government use. Located on mountain tops and other high locations, most duplex repeaters have the following attributes:

- **Output Frequency:** When using repeater lists to find what is available in an area, the output frequencies are listed. When tuned to a designated output frequency, you will hear the repeater's outbound traffic. For sending message to the repeater for rebroadcast, you will use what is called an "input frequency offset."

- **Input Frequency Offset:** Duplex repeaters use an input offset frequency to receive message for rebroadcast and the offset frequency will be just above (positive offset) or just below (negative offset) the assigned output frequency.

  For example, in the US the usual offset for Amateur transmit frequencies in the 144 MHz 2 M band is 600 kHz and 5 MHz in the 430 MHz 70 cm band.

- **Input Frequency Private Line (PL Tone):** A PL tone is one of several types of sub-audible codes used to prevent a repeater from responding to unwanted signals or interference. It is often used by duplex repeaters to prevent unwanted transmission.

  A PL tone is configured by the repeater operator and operators that use the repeater must configure their transceivers to begin each transmission on the assigned frequency with the designated PL tone. If the tone feature is not enabled or it is set incorrectly on the sending transceiver, the repeater will ignore those incoming messages until it detects the assigned PL tone.

For global tribulation survival applications, the pros and cons of duplex repeaters are:

- **Duplex PRO:** Duplex repeaters will give your survival community a huge edge in maintaining communications inside and outside of your perimeters.

- **Duplex CON:** Duplex repeaters require sophisticated controllers and must be configured and maintained by competent technicians with advanced test equipment. Furthermore, duplex repeaters are normally located in fixed locations and will require ongoing supervision and maintenance by your technical team.

When using printed or online duplex repeater listings, you will typically see three things. For example let's assume you're working a duplex repeater that is listed with the following specification: "145.150, 123.0, -." This translates to an output frequency of 145.150 MHz, a PL tone frequency of 123 Hz, and a negative offset (-600 kHz).

Now for a final note on duplex repeaters during the global tribulation, the one duplex repeater feature your survival community should avoid is the PL tone and for two good reasons.

By law you cannot transmit encrypted or scrambled messages, so everything is "in-the-clear" as HAMs say, so security is one issue. The other is that requiring a PL tone does not stop anyone from transmitting on an assigned duplex repeater frequency. What this means is that you cannot hear them but they can hear you and what you're saying. The other reason is interference caused by malicious frequency jamming.

Duplex repeaters used in metropolitan areas serve large numbers of HAM operators and by requiring a PL tone, they eliminate unwanted signals and interference from overlapping traffic. However, during a global tribulation, your repeaters will be serving a much smaller population of operators. Besides, there is another type of repeater you will find most useful during the coming global tribulation. It's called a cross-band repeater.

Cross-Band Repeaters

Cross-band repeaters are sometimes used by amateur operators as a means to extend the range of the HT transceivers by using a dual-band, cross band capable base or mobile transceiver in a central location.

During the global tribulation, your community will likely use cross-band repeaters the most, assuming that you've chosen the right kind of dual-band mobile transceivers, ones capable of working as true cross-band repeaters.

Hence, when it comes to setting up **cross-band** repeaters, the operative word is "true" so let's take a close look at what "true" really means in this sense.

- **"True" Dual-Band Transceiver:** A true full-duplex, dual-band transceiver that can simultaneously receive on two different frequencies and can be on the same or different bands.

  When a manufacturer claims their transceiver offers true dual-band functionality, you'll need to see something like "True Dual Receive (VHF+VHF/VHF+UHF/UHF+UHF) operation." This tells you that you can simultaneously monitor two frequencies on the VHF band or the UHF band. Or, you can monitor one frequency on the VHF band and another on the UHF band.

- **"True" Cross-Band Repeater:** Mobile transceivers that offer "true" cross-band functionality are typically "true" dual band transceivers that also offer "true" full duplex cross-band repeater functionality. So, when it comes to looking for a suitable transceiver for use as a cross-band repeater, remember to ask yourself, "is it a true-true?"

Please remember one simple rule. Whatever you do, it has to be a "true-true" solution straight out the box without the need for any geeky work-arounds.

To illustrate this, let's see how true cross-band repeater functionality can be used to support a field team operating beyond your outer perimeter.

When organizing a cross-band receiver in support your field teams, you need to define three elements. The transceiver to be used as a cross-band repeater, where it is located and the types of transceivers it serves and on which bands.

For the purpose of this discussion we establish those three elements as:

- **Cross-Band Repeater:** The Kenwood TM-V71A 50-watt mobile transceiver, offers out of the box functionally as a "true true" cross-band repeater.

- **Repeater Site:** A mountain ridge location inside your outer perimeter with a clear line-of-sight to the area your field team is operating in and your command base station.

- **Base Stations:** Operators working in fixed locations such as your command base station and operators in the field who will access the repeater with a suitable transceiver.

As to your command base station, this will be discussed in greater depth later on. For now, suffice it to say that the Kenwood TM-V71A 50-watt mobile transceiver with true, built-in cross-band repeater functionality will be the line-of-sight workhorse for your command base station and associated repeater sites.

For your operators in the field, we previously introduced two amateur HT transceivers. The single band Kenwood TH-K20A VHF HT and the dual-band Yaesu FT-60R VHF/UHF HT.

Out of the box, both are designed to work with simplex and duplex repeaters with input and output frequencies in the VHF band. The only significant difference between the two for the purpose of this discussion is that the Yaesu FT-60R can work repeaters in the UHF band as well. Likewise, both offers a PL tone function for duplex repeaters in the VHF band as well.

Therefore, the TH-K20A is ideal for bi-directional communication via a true cross-band repeater using the VHF band. Because a PL tone is not used, all that is needed is a VHF frequency such as 146.50 MHz. Your field operative will configure their TH-K20A HT transceivers to this frequency.

On the other hand, your command base station will work the other side of the cross-band repeater with another Kenwood TM-V71A on the UHF band.

At first glance, you may be feeling a serious twitch of complexity here. So, let's dispel that notion by seeing how straightforward the process of setting up a cross-band repeater system for your survival community can be when you use the right equipment.

Cross-Band Repeater Setup

As an example let's position our cross-band repeater on a ridgeline. Your survival community is on one side and your security team is beyond the outer perimeter on the other side. To get started, your technical team will use a Kenwood TM-V71A 50-watt mobile transceiver with true, built-in cross-band repeater functionality.

All of the following steps can be performed in the field without additional testing equipment or devices except those necessary to operate the repeater.

1. **Repeater Installation Site:** The process begins by installing the TM-V71A on a ridgeline with a clear line-of-sight view of the operational area for your security team outside the outer perimeter. On the reverse side, you should see the antennas used by your command base station.

2. **Setting Up the Outside the Perimeter Frequency and Power:** For your field teams, the TM-V71A is configured on the VHF side. The repeater is set to 146.50 MHz which your security team will use with their single-band VHF TH-K20A HT's.

   Initially, the transmit power for the VHF frequency should be set to maximum for the greatest possible range, and then lowered as circumstances allow. This is in keeping with one of amateur radio's cardinal rules: Always use the lowest possible transmit power.

   Note: NEVER transmit on a transceiver without it being connected to a properly configured antenna.

3. **Setting Up the Inside the Perimeter Frequency:** On the UHF side, the TM-V71A cross-band repeater is configured to use with UHF frequency 443.50 MHz. Since

the distance between the ridgeline and your command base station will likely be substantially less, the UHF side of the repeater should be configured with the lowest possible transmit power setting. Be sure to test this.

4. **Secure DTMF Cross-Band Repeater Control:** Amateurs use Dual-tone multi-frequency signaling (DTMF) to remotely control their transceivers. DTMF is a common feature that works in a similar way to the keypad on your telephone. This functionality is either built into the hand mike or the transceiver itself.

   In this example, the UHF side of the cross-band transceiver will be used to remotely control it using DTMF commands issued by your command base station. This is because the FCC only allows DTMF control codes to be transmitted on the 440MHz UHF band.

   Note: You can also secure your DTMF communications, with a remote access password.

5. **Setup Auto Announce:** Until the global tribulation is in full swing, FCC repeater compliance is a must. Therefore, your repeater needs to identify itself every ten minutes while on the air.

   The TM-V71A offers with an auto announce feature, which can be configured to use Morse code (or voice), to automatically announce your call sign every ten minutes. It is advisable to configure the TM-V71A to transmit call the sign in Morse code.

6. **Reduce Power Where Possible:** The TM-V71A offers a bright, backlit display and control tones. However, operating the TM-V71A in cross-repeater mode requires more power than normal and so expending power on these user features is unnecessary. They should be turned off or down to conserve power wherever possible.

Your TM-V71A is now ready to support your field team beyond the outer perimeter. Once it is powered up and operating, you may start using it.

When you do, the communication path between the security team and the TM-V71A repeater will be on the VHF side of the repeater.

On the UHF side, the path will be between your TM-V71A repeater and your command base station which will also serve as the DTMF remote access control for the repeater.

Another benefit of using **cross-band** repeaters such as the example configuration discussed above is "daisy chains". Used in a daisy chain configuration, full-duplex **cross-band** repeaters can be daisy-chained with alternating bands and frequencies configurations. The result is a practical way to communicate across multiple ridgelines.

In addition to this, you can use a more elaborate cross-band repeater antenna setup to further increase range and enhance privacy. Let's see how that works.

Repeater Antennas

In the previous chapter we mentioned two different kinds of wire antennas for use with QRP CW HF transceivers: end fed and dipole.

An end fed antenna when mounted vertically is omnidirectional, which means that it transmits in a 360° beam width with equal strength. Other examples of omnidirectional antennas are the rubber duck antennas held vertical, supplied by the factory for both the Kenwood TH-K20A HT and the Yaesu FT-60R handy-talkies discussed previously.

Likewise, the 5/8 wave whip antenna previously recommended for the TM-281A VHF mobile transceiver is also an omnidirectional antenna. So, what is the difference between omnidirectional antennas and **directional antenna**s? A directional antenna like a horizontal dipole transmits in a much smaller, more focused beam width pattern.

Ergo, if you're just looking to contact someone without knowing their physical location, you want to use an omnidirectional antenna. However, if you know where your contact is and you want to be able to transmit with less power and with greater signal security, a directional antenna is what you need. Here you will need to implement a well-defined signal control strategy using directional antennas which offer two advantages, a narrow beam width and reduced transmit power.

Signal Control Strategy

Signal control strategy involves three things that will help to ensure the privacy of your communications:

- **Signal Intelligence:** In a signal control strategy the only form of signal control for omnidirectional antennas is signal intelligence. The issue is, that these antennas transmit a 360⊕ beam width with equal power in every direction. Therefore, detailed and specific information about your community's members, location, etc. should be treated as classified.

- **Directional Antennas:** With directional antennas, you need to know who you want to communicate with and where they are located. This is good, because when using a directional antenna, unwelcome third parties seeking to eavesdrop, will find it difficult or impossible to do so if they are outside of your antenna's beam width.

- **Reduced Transmit Power:** When you point a directional antenna with its narrow beam width, less transmit power can be used than with a 360⊕ beam width omnidirectional antenna. Depending on the type of directional antenna used and how it is configured, considerable power savings can be enjoyed.

No matter whom you are communicating with, where they are located, and regardless of the type of antenna your operator is using, the bottom line with a signal control strategy is that there must be guidelines and they must be enforced. Doing so will reduce the possibility of transmitting sensitive information to unintentional third parties.

Now that you know the benefits of a directional antenna, let's discuss a very popular type of directional antenna, called a "Yagi-Uda" or more commonly, a "Yagi" named in honor of the two Japanese HAM operators who invented the design.

Yagi Directional Antennas

A Yagi antenna is a directional antenna composed of multiple parallel elements (metal rods) in a line. With a 50° to 70° beam width, they're well-suited to receiving lower strength signals. However, a Yagi can still send and receive signals outside of its intended beam width, but in a very limited way, via its side and rear lobes. Nonetheless, eavesdroppers working outside of your beam would need to be relatively close to intercept your communications if you are using minimum power policies.

OMG, was that geeky or what?

Better yet, let's start over with a simple question. "How can I recognize a Yagi antenna?"

That's easy, if you have gray hair. If not ask a grandparent to think back to the days when folks mounted antennas on the roof tops of their homes during the early days of television. In most cases they would be using a directional antenna like a Yagi.

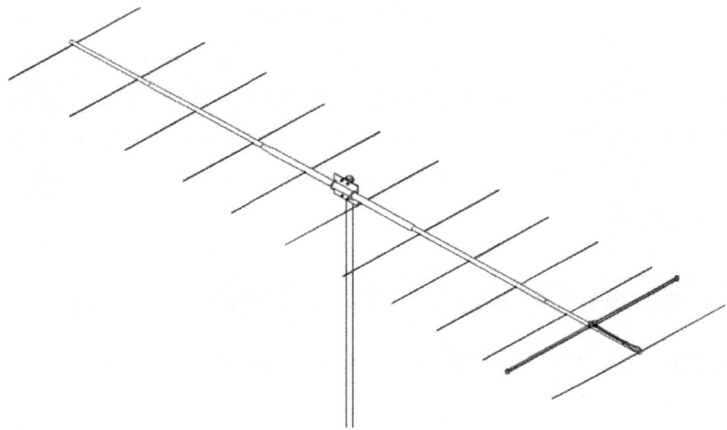

For early television viewers, the benefit of a Yagi directional antenna was that it would improve reception when pointed at a broadcast tower used by the local television stations.

If multiple broadcast towers served your area from different directions, the base of the antenna could be mounted on an all-weather powered rotor; the antenna's aiming direction could then be controlled from inside the home.

With this in mind let's use multiple Yagi antennas and minimum power with our cross-band repeater system to enhance signal control.

Cross-Band Repeater Yagi Antenna Setup

Directional antennas such as the popular Yagi are relatively inexpensive, easy to transport (when folded) and can be very helpful in supporting field teams operating beyond your outer perimeter and with your repeater network.

To illustrate these benefits, let's assume your field teams are equipped with a foldable, handheld, 3-element Yagi antenna. Let's also assume that your technical team is supporting your field team by setting up a cross-band repeater with an unobstructed line-of-sight to your field team and to your command base station. In this case, your technical team will use more robust, 13-element Yagi antennas.

So, what is the difference between 3 and 13-element Yagi antennas? Gain and Portability.

- **Gain:** A Yagi is a high gain antenna which means it will increase the transmit range of a transceiver on a given frequency. This is expressed in the term decibels (dB), and with a Yagi the more elements you add, the greater the dB gain.

  The gain with a 3 element Yagi is approximately 3 to 6 dB, and as a rule of thumb, every 3 dB of gain equals a doubling of the transmit power's effectiveness. Ergo, a field operator using a VHF transceiver with a 3-element Yagi, with a modest 3 dB gain, will double their effective send and receive range.

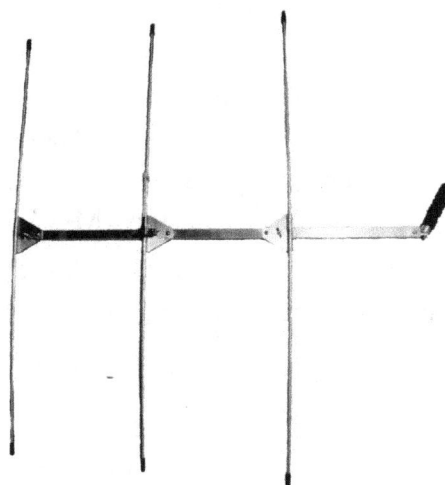

  On the repeater side, 13 element Yagis are used because thirteen elements gives the antenna yield a gain of 9 to 10 dB or more. The receive and transmit power of the repeater station are therefore effectively quadrupled, thereby allowing lower transmit power settings for enhanced signal control.

- **Portability:** Handheld Yagi antennas are a common favorite of HAMs for backpack field operations. With 3 elements a VHF Yagi can easily be folded up into a small carry bag. The operator can quickly unfold the antenna and hold it much like a pistol while aiming it at the repeater.

  On the other hand a 13-element Yagi is unwieldy for handheld operation and must be mounted to an antenna mast or camera tripod.

At this point, what have we got? On the one side of the repeater, a field operator is working beyond your outer perimeter and they are equipped with a foldable 3-element Yagi handheld antenna. On the other side, your command base station has an antenna farm designed to work multiple bands and services.

On a mountain ridge between the two is where your technical team will setup a cross-band repeater using a TM-V71A and two, mounted, 13-element Yagi antennas.

When briefing your technical team for this mission, they will need to know if they are to prepare a temporary or fixed repeater installation.

If the two Yagi antennas are fixed, they can be mounted to the roof of a ridgeline observation post or on a pole. If temporary, they may decide to use a camera tripod for each antenna instead. Either way, how the antennas are connected and aimed remains the same, so let's see how that works:

- **2 M Yagi Pointed to Field Team:** A 13-element Yagi 2M VHF directional antenna is aimed from the ridgeline out towards the area where your team will be operating with single-band VHF HT transceivers beyond your outer perimeter.

- **70 cm Yagi Pointed to Command Base Station:** A 13-element Yagi 70cm UHF directional antenna is aimed from the ridgeline back towards the antenna farm serving your command base station. Inside the station, another TM-V71A is installed.

- **Connecting to the TM-V71A Cross-Band Repeater:** The TM-V71A has only one antenna connector. Therefore, the coax cables from the 2M and 70cm Yagi antennas are joined using a two-way splitter, which has three connectors. Two for the antennas and one is for the TM-V71A repeater.

After setting up the two directional antennas and connecting them to the TM-V71A, your team performs radio checks in both directions to align both Yagi antennas.

Once this is completed, you are ready for cross-band repeater operations with the following benefits:

- **Increased Directional Range:** If your transmitter's maximum power is 50 watts, then an omnidirectional antenna will propagate that power in all directions equally. However, when you aim that same 50 watts with a directional antenna like a Yagi, the narrower beam width will channel the transmit signals within a more restricted range of coverage.

- **Reduced Directional Power:** For the sake of discussion, let's assume that a Yagi antenna is able to increase the range available with 50 watts of power by double. If that extra range is needed, you would obviously stay with the 50 watt power setting. However, if that extra range is not required, you can reduce the power by half to 25 watts. This would mean lower power consumption and fewer unintended receivers able to listen to your traffic.

- **Enhanced Directional Privacy:** With reduced directional power, you reduce the number of unintended receivers operating beyond the frontal area of your field team, but this is only half of the benefit. The other is that you will also limit unintended receivers from receiving your traffic that are operating to either side and along the flanks of your field operatives.

When you take into consideration the ease of setup, a dual-band transceiver like the Kenwood TM-V71A with "true true" full-duplex cross-band repeater capability offers you a more reliable way to direct the movements of their field teams operating outside the outer perimeter with less power and less eavesdropping.

Now, what about the autonomous (undirected) teams or individuals? In this case using a daisy chain of **cross-band** repeaters to work across multiple ridgelines may not always be a good or desirable option.

In this case, there is a different way to work across the ridgelines without repeaters and it's powerful. Namely, High Frequency (HF) field transceivers.

11

HF Field Transceivers

When you have autonomous (undirected) teams or individuals working areas beyond the outer perimeter outside of the range of your line-of-sight repeater system, you will need to implement a high frequency (HF) strategy.

The DX advantage of HF transceivers is due to the inherent ground wave and sky wave propagation characteristics of HF. In this case, the goal is not to make overseas DX contacts, it is to allow you and your autonomous field teams to communicate across multiple ridgelines with low-power QRP HF transceivers via phone and CW modes. For phone mode, your minimal target range is 10-20 miles and for CW mode, it is 20-50 miles.

Previously, we discussed two QRP HF transceivers for use with your level 1 of your community communications strategy: SSB CB (phone mode only) and QRP CW 40m (CW – Morse Code only.)

For level 2 field operations, a CW work-around can give an SSB CB both phone and CW modes with passable performance. The inexpensive MFJ 9040 discussed in the previous part of the book is a CW-only transceiver. The MFJ 9040 is an effective CW-only emergency transceiver and it is advisable that field teams carry one as a standard issue item.

However, because phone mode is necessary for level 2 of your community communications strategy, an upgraded model, the MFJ 9440 will be discussed in this chapter. It comes with phone mode as the standard mode in the base model version and CW with keyer support can be added using an optional adapter card.

With this in mind, here are the three QRP phone and CW options that will be discussed in this chapter to help illustrate three solutions which are based on desired functionality and budgeting concerns.

HF Low-power Transceiver Description	CW	Phone	Data	Watts
Level 1 – SSB CB 11m w/Code Oscillator and Key	•	•	n/a	12
Level 2 – QRP HF 40m Backpack CW & Phone	•	•	•	10
Level 2 – SOTA 80-10m HF CW & Phone	•	•	•	10

If your community is working on a tight budget, the first level 2 field radio to consider for use is a SSB CB mobile transceiver with a code oscillator and key.

SSB CB 11m w/Code Oscillator and Key

SSB CB is the easiest and least expensive way to obtain the HF DX benefits for phone mode. This channelized service resides on the HF 11m band, at the high end of the HF spectrum.

Previously, we discussed the Uniden Bearcat 980SSB SSB CB. When configured as a backpack radio, the low power consumption this transceiver makes it a good choice with a 12 V gel cell or Li-Ion battery.

Bearcat 980SSB CB 11m			
Input Power (VDC)	Current Drain Amps (A) Milliamps (mA)		Transmit (Watts)
VDC Input Range	Receive Drain	Max Transmit Drain	Maximum Output
13.8 V DC	650 mA	4 A	12

When used in the field, the 980SSB offers four important benefits:

- 40 AM CB Channels Plus SSB – Includes Upper and Lower Sideband channels.
- 7-Color LCD Display – Provides easy viewing in any light condition.
- S/RF/SWR Meter – Makes it easy to tune your CB antenna for optimum performance.
- Ergonomic Layout – The controls are well laid out and the digital display is large.

However, one concern with the 980SSB is the form factor, for backpack configurations. To illustrate the difference let's compare the 980SSB with the Kenwood TM-V71A.

Radio	Width	Height	Depth	Weight
980SSB	7.25" (184 mm)	2.25" (57.15 mm)	7.50" (190.5 mm)	2.2 lbs (1 kg)
TM-V71A	5.51" (140 mm)	1.69" (43 mm)	8.39" (213.1 mm)	3.3 lbs (1.5 kg)

While the 980SSB weights a pound less than the TM-V71A, its bulkier form factor will be a consideration. The 980SSB costs about a third of the TM-V71A's cost and it does not need a repeater to send and receive signals across multiple ridgelines. For a survival community on a tight budget, this is as good as it gets.

This being said, the one thing that the 980SSB and the TM-V71A have in common is that neither supports CW mode. Both lack a CW key port which is typical for these types of radios. This is not to say it cannot be done. In fact, there is a simple way to send Morse code messages using any type of transceiver offering phone mode.

Sending Morse Code with Any Transmitter

CW mode has the longest-working range of any mode because it is simple and efficient. However, there is a phone mode work-around that will enable any transceiver to be used to send Morse code.

To send CW using any phone mode transceiver with a microphone, you'll use a Morse code training tool called a code oscillator with a telegraph key.

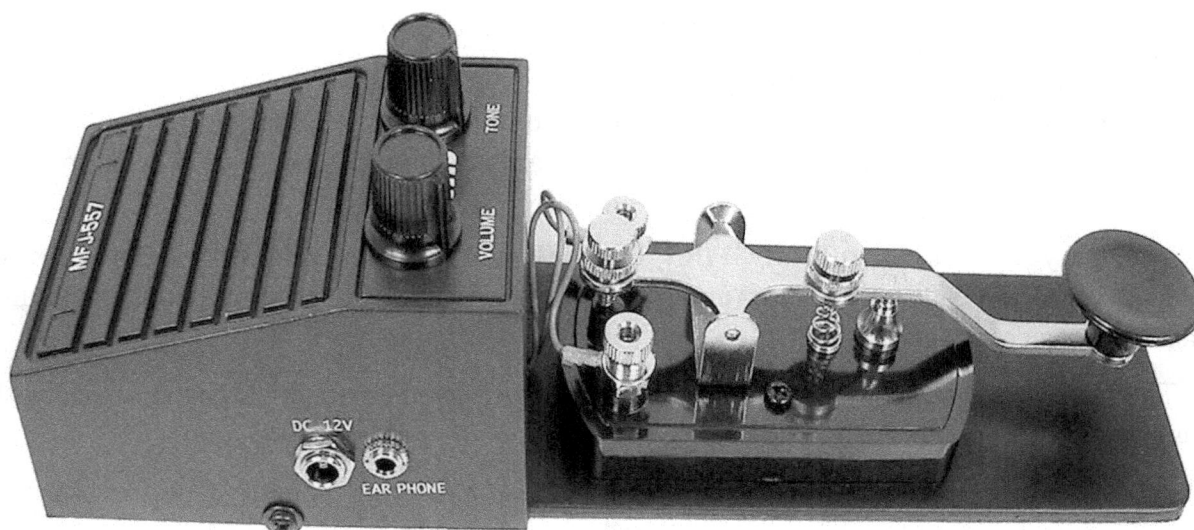

Here we need to distinguish between two CW mode terms, sidetone and code oscillator as both devices allow a CW operator to hear what they're keying, as they key it.

- **Sidetone:** A feature built into a CW enabled transceiver that provides real-time audible feedback to operators as they key their messages.

- **Code Oscillator:** A stand-alone training tool, a code oscillator emulates a sidetone so Morse code students hear what they're keying without being connected to a transceiver.

An example of a suitable Morse code oscillator training tool is the MFJ-557, which features a Morse Code straight key. Powered by a 9 V battery, it can be used in the field, but you can construct small, efficient, do-it-yourself (DIY) code oscillators with just a few basic parts such as a home door buzzer.

How effective is this workaround? That's difficult to say because Morse code messages keyed in CW mode will go further than Morse messages transmitted via this Morse code phone mode work-around. How much further is very difficult to say, but one way to measure the difference is with is signal readability, where a readable signal is one you can clearly hear and understand.

When the other party reports that your spoken voice phone mode signal can be heard but is unreadable, this is when you switch to the Morse code phone mode work around. In this case, the simple dit and dah tones should be readable as a coherent message.

The work-around process for sending Morse code with a microphone, code oscillator and key is as simple as 1-2-3.

- **Start Transmit Mode:** To begin, hold the microphone over the code oscillator speaker, then press and hold the push-to-talk (PTT) transmit key on the microphone to initiate voice mode transmit.

- **Key the Message:** Then begin keying the message with the key (or keyer) connected to the code oscillator. The transceiver will then send the code oscillator tones as a phone mode message.

- **Return to Receive Mode:** When finished transmitting the message, release the PTT key on the microphone to put the transceiver back into receive mode so that you can hear the answer to your message.

Whether you are using an FRS, GMRS, UHF, VHF, or CB radio, this work-around process for sending Morse code will work, provided the code oscillator generates a clear tone with distinct Morse code dits and dahs.

There are two important things you need to remember about how a microphone is used with this simple work-around.

- **No Powered Microphones:** Powered microphones are a popular accessory for CB enthusiasts, but can cause overmodulation. If you are using a powered mic, disable this feature or turn it all the way down. Better yet, use the simple, unpowered dynamic microphone that came with your transceiver.

- **Microphone Orientation:** Proper microphone orientation means that you hold the microphone at an angle perpendicular to your face. This provides a clear tone without distortions because the sound, whether it be your voice or the dits and dahs of a Morse code oscillator workaround, are directed toward the pickup-pattern of the microphone.

The bottom line is that your field teams will each need a dedicated radio operator who is highly proficient in Morse code. Likewise, all of the operators in your command base station will need to know Morse code, when there is too much interference for voice.

When there is too much interference for voice communications, switching to this work around will, in most cases, be how you get the message through. A critical message such as "a warlord and 200 thugs are headed your way," will get through when it needs to get through.

Not only is it a good idea to equip all of your field teams with Morse code oscillators and keys, your technical team can also design specialized cables to connect a telegraph directly into the microphone connector port on your transceiver.

To illustrate how this is done, let's use a Uniden Bearcat 980SSB. In this example, a powered code oscillator with a 3.5mm audio output jack is connected via a special cable to the 980SSB's microphone port. You could then send Morse code signals with a key instead of a microphone.

To hear incoming messages without being overhead, your radio operator connects ear buds or a headset to the 980SSB's audio output port. Then, by enabling the 980SSB's talkback mode you can hear your own Morse code transmissions just as you would with a CW transceiver with sidetone capability. This way, you can hear both the incoming receive signals and the Morse code you are transmitting.

The bottom line is that an SSB CB is the least expensive way to hop across the ridgelines, provided you understand how to work within the limitations of this consumer service 11m HF QRP transceiver. This is not to say it is the best option. It is not. But if you're working on a very small budget, it is passable.

A better next option is a single band tunable QRP 40m transceiver with built-in support for Phone and CW modes, such as the MFJ-9440. While this options cost more than the SSB CB option discussed above, the benefits are substantial.

MFJ-9440 40m Transceiver

In the previous chapter, we introduced the MFJ-9040. An inexpensive, tunable QRP 40m CW-only transceiver that is suitable for level 1 emergency communications. While this is a good

backup or emergency transceiver, is not truly suitable for day-to-day missions your field teams will be conducting while operating beyond your outer perimeter

However, there is a similar model by the same manufacturer that is a good level 2 field operations option because it offers both phone and CW modes. It is the MFJ-9440 with the CW adapter add-on.

Although it costs approximately a third more than the MFJ-9040, the MFJ-9440 offers twice the transmit power, SSB phone mode and key and keyer support.

Aside from the SSB phone mode, the big difference between the two is power as seen in the tables below.

MFJ-9040 Level 1 – Emergency CW Transceiver			
Input Power (VDC)	**Current Drain Amps (A) Milliamps (mA)**		**Transmit (Watts)**
VDC Input Range	**Receive Drain**	**Max Transmit Drain**	**Maximum Output**
12 to 15 V DC	50 mA	1.2 A	5

MFJ-9440 Level 2 – SSB CW Field Operations Transceiver			
Input Power (VDC)	**Current Drain Amps (A) Milliamps (mA)**		**Transmit (Watts)**
VDC Input Range	**Receive Drain**	**Max Transmit Drain**	**Maximum Output**
12 to 15 V DC	100 mA	2.2 A	10

Out of the box the MFJ-9440 only supports phone mode in the upper part of the 40m band, from 7.150 MHz to 7.300 MHz. This sub-band is designed for phone and data modes and is only available to licensed HAMs with a General Class License or higher.

For CW Morse code, a special add-on circuit CW card, the MFJ-415B must be installed in the MFJ-9440 SSB transceiver. The MFJ-415B card does the following:

- **CW Frequencies:** The lower part of the 40m band (7.000 MHz to 7.150 MHz) is designated for use by CW and is available to licensed HAMs with a Technician class license or higher.

- **Key and Keyer Support:** When equipped with the MFJ-415B, the MFJ-9440 offers two ports. One for a straight key and another for Iambic dual-paddle keyer operators. Both feature sidetone.

In terms of field operations, two advantages of the MFJ-9440 that help to justify its higher cost are:

- **Calibrated S-Meter:** Like the 980SSB, the MFJ-9440 sports a useful S-Meter that makes it easier to peak the tuner or to find the best beam heading.

● **Rugged Transmitter:** The MFJ-9440 is designed to tolerate a higher level of mismatch between an antenna and the feed line that connect to it than most other radios which can be damaged at the lower SWR level. This is monitored with the S-Meter.

These two benefits will be of special value to field team operators when moving between different areas in the course of a mission.

The bottom line with a modestly affordable QRP phone/data transceiver such as the MFJ-9440 is that it has one limitation in common with an SSB CB, namely, access to only one HF band. Is this acceptable? Who can say? During the global tribulation, natural events will create a fluid and changeable operating environment in which any one band can become unusable for an unforeseen period of time.

For this reason, the ability to operate on multiple HF bands will add an extra layer of usefulness that could spell the difference between life or death in mission-critical situations.

Therefore, if your budget can allow you to work with multi-band, multi-mode portable backpack QRP transceiver designed for use in the field, such as the Yaesu FT-817ND.

FT-817ND Portable Backpack Transceiver

First introduced in 2004, the Yaesu FT-817 was the world's first HF/VHF/UHF self-contained, battery-powered, multi-mode SOTA-class portable transceiver. It is an affordable backpack transceiver and typically used for outings in regional areas. Based on customer feedback, the FT-817 was upgraded to the FT-817ND in 2017.

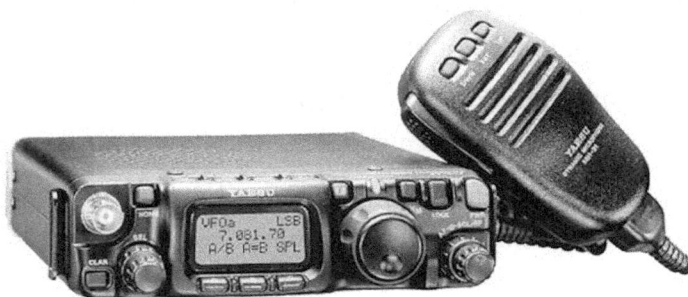

In 2018 Yaesu released the FT-818, as a replacement for the FT-817ND. It is the same basic transceiver with new features desired by many existing FT-817ND owners. A minor concern is that the FT-818 has yet to build a history of reliability such as the models it replaces. Therefore, it is still somewhat of an unknown.

A well-regarded 5-watt backpack transceiver, the Yaesu FT-817ND supports multiple modes: CW, Phone, USB, Data, Packet and covers the amateur frequencies for HF/50/144/430. It also supports that 5.1675 MHz Alaska Emergency Frequency (USA only) which resides between the 80m and 60m bands.

At the time of its introduction, military surplus was the primary source of backpacker radios and its continued sales today stand as a testament to Yaesu's marketing plan for this model. Nonetheless, there were teething pains, which is something to expect with any new RF product.

Yaesu FT-817ND Transceiver			
Input Power (VDC)	**Current Drain Amps (A)** **Milliamps (mA)**		**Transmit (Watts)**
VDC Input Range	**Receive Drain**	**Max Transmit Drain**	**Maximum Output**
12 to 15 V DC	450 mA	2 A	5

The problem with the original FT-817 models was that they were highly susceptible to damage from mismatched antennas. This deficiency became a real issue for backpacking HAMs because, unlike more modern designs, the FT-817 lacks a built-in antenna tuner though there are aftermarket components and external antenna tuners.

For this reason, Yaesu introduced its ATAS-25 antenna system. Designed specifically for the FT-817ND, it which supports the: 40m, 20m, 15m, 10m, 6m, 2m, 70cm amateur bands.

For survival mission roles for field teams operating beyond your outer perimeter, it is highly advisable that a field team equipped with this portable transceiver be likewise equipped with the ATAS-25 antenna system.

Unlike electronic internal and external antenna tuners, the ATAS-25 is a manual antenna tuning system. When changing bands, there are tuning controls on the actual antenna assembly which are used to rough tune and then fine tune the antenna without the need for an external antenna tuner.

For autonomous teams and individuals operating beyond the range of your repeater network, there is one drawback to this older design – 5 watts of transmit power.

While 5 watts is enough power when operating the FT-817ND in a normal environment, for field operations in the global tribulation environment, 10 watts with a more current technology design and mission role may be preferable. In this case, you want to move from multi-mode, multi-band QRP backpack portable transceivers to *Summits on the Air* (SOTA) class transceivers.

Summits on the Air (SOTA)

SOTA stands for Summits on the Air. SOTA is an award scheme for radio amateurs that encourages portable operation in mountainous areas and is fully operational in almost one hundred countries worldwide.

First launched in Great-Britain in 2002, SOTA has grown into a highly respected and popular, worldwide program. It no doubt helped inspire the development and introduction of the Yaesu FT-817 in 2004.

Each SOTA member country has its own association which defines the recognized SOTA summits for that Association. Members of these programs earn credits as activators by transmitting messages from summits or as chasers receiving those messages. Awards are given and certificates are available for various scores. In addition, their achievements are recorded in a SOTA online database.

To participate in the SOTA program:

- You must be a licensed HAM amateur operator.
- You cannot operate in close proximity to a car.
- You must travel to your location on foot or by bicycle.
- You must physically carry all equipment you will use.
- All equipment must be powered by portable sources such as batteries and solar cells. No fuel-powered generators.
- The operating area must be within one topographical map contour line (approximately 25 meters) below the summit.

HAMS that participate in the SOTA program favor more expensive handheld QRP HF transceivers, designed for the rigorous demands of the SOTA environment, such as the Elecraft KX2. And yes, it's a lot more expensive so get out your Backgammon doubling cube.

Elecraft KX2 SOTA-Class Transceiver

Elecraft is an established American manufacturer. Founded in 1998, they've found a niche with high-performance transceivers and accessories that HAMs can purchase as kits or as factory-assembled products.

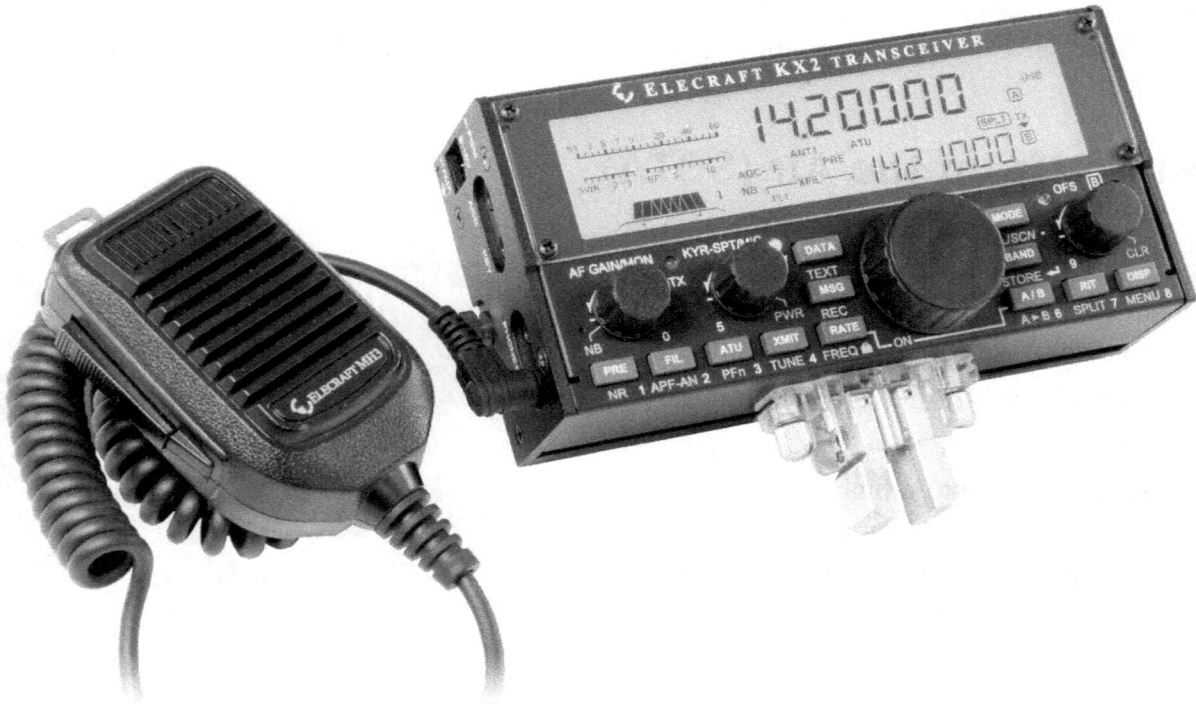

A recent addition to their product family is the 10 watt KX2 SOTA-class multi-band HF transceiver for nine HF bands, from 80m to 10m.

Who can afford a KX2? Typically, it's the HAMs with deep pockets. For them the factory assembled KX2 is the Rolls Royce of SOTA handheld transceivers and priced about the same. If you can swing the cost of this transceiver in your budget, this handheld SOTA-class transceiver has more than just drool factor, it's got the goods.

Elecraft 80-10 M KX2 Transceiver			
Input Power (VDC)	**Current Drain Amps (A)** **Milliamps (mA)**		**Transmit (Watts)**
VDC Input Range	**Receive Drain**	**Max Transmit Drain**	**Maximum Output**
12 to 15 V DC	135 mA	2 A	10 (80 – 10 M)

Let's compare the KX2 with the FT-817ND and the MFJ-9440.

KX2 vs FT-817ND

Multiple HF bands mean that with either radio, your field operatives will have more options for sending signals during adverse periods of global environmental interference.

In a head-to-head comparison the FT-817ND offers a more robust range of bands including VHF and UHF. However, the nine HF bands available with the KX2 represent the bulk of the workhorse HF bands.

Also keep in mind that while the Yaesu FT-817ND supports the amateur frequencies on the line-of-sight VHF and UHF bands, it has the same transmit power as a Yaesu FT-60R handy-talkie previously discussed.

For line-of-sight the advantage is minimal compared to more powerful dual-band VHF/UHF mobile transceivers such as the Kenwood **TM-V71**A as previously discussed.

However, a serious concern with the FT-817ND is that it is not available with a built-in automatic antenna tuner option from the factory.

However, Elecraft does offer a built-in automatic antenna tuner option for the KX2. It is the KXAT2 automatic antenna tuner and it gives the KX2 a wider range of antenna choices.

KX2 vs. MFJ9440

As with the singe band MFJ-9440, the multiple band KX2 also offers CW and SSB phone mode. While both transceivers offer 10 watts of transmit power, the KX2 has one additional mode of extreme value for your field operatives – data.

Your operatives with data mode can take a picture with a small tablet, and via a USB connection to the radio, transmit a graphic file back to your command base station.

KX2 vs. FT-817ND and MFJ9440

When comparing the KX2 with both the FT-817ND and the MFJ-9440, there are new technologies that put the KX2 well ahead.

- **Advanced CMOS Sensors:** Newer radios like the KX2 have more advanced CMOS technology with sensors designed to prevent transistor damage.

- **Internal, Wide-Range Automatic Antenna Tuner:** The KX2 is available with the optional internal KXAT2 automatic antenna tuner. While many HAMs prefer stand-alone antenna tuners for multi-band HF transceivers, the military standard is preferable.

  Military field radios are equipped with built-in automatic antenna tuners to give the best possible signal while protecting the radio's transmitter from damage resulting from a mismatched antenna.

- **Software-defined-radio (SDR) Architecture:** Newest generation radios are designed to be software-programmable (at the factory) to give the radio more features and functions. With SDR, more general purpose components can be used to simultaneously reduce the complexity of the radio's electronic circuits by reducing the number of parts necessary.

SDR is a huge evolutionary step forward for the radio industry and this is something you want to consider during your evaluations, because this helped Elecraft to reduce the size of the KX2. A difference that truly stands out when you compare the form factor differences between the FT-817ND and the KX2 are:

Radio	Width	Height	Depth	Weight
FT-817ND	5.31" (135 mm)	1.5" (38 mm)	6.50" (165 mm)	41 oz (1.17 kg)
KX2	5.8" (147.3 mm)	2.8" (71.12 mm)	1.5" (38 mm)	13 oz (0.37 g)

As you can see in this form factor comparison, the KX2 offers twice as much transmit power with a fraction of the size and weight of the older technology FT-817ND. Like they say, "you get what you pay for," but what has that got to with budgeting? Life itself.

The Final Buying Decision

Now let's talk heart to heart. Sending someone into harm's way is not a hypothetical situation. These kinds of risks are going to be taken because there is something important that needs doing. It could be for long-range scavenging for badly needed medicine or equipment, or to surveil a warlord harboring evil designs upon your community. Whatever it is, precious lives will be at stake both in the field beyond the range of your mountain top repeaters and within your community itself.

Here is where you and your technical team will have two basic choices to consider. Save the money or honor the risk.

- **Save the Money:** The Bearcat 980SSB SSB CB and the MFJ-9440 SSB discussed above represent the most affordable options available to you. However, when you begin looking at HF transceivers capable of multiple bands and modes, you might as well throw a backgammon doubling cube to calculate the added costs.

- **Honor the Risk:** Costs for equipment designated for use with your community communication strategy, can be controlled with rationing and reduced functionality. However when good people are far afield and putting their lives on the line, here is one time when you should spend whatever is necessary and honor their risk – if possible.

If your only goal is to save the money, you already have two excellent low budget options, the Bearcat 980SSB SSB CB w/code oscillator and key and the MFJ-9440 SSB with the MFJ-415B add-on CW circuit board.

If you can afford to spend more for a backpack multi-mode, multi-band portable transceiver the FT-817ND is an old, tried-and-true design for transmitting an HF/50 signal over multiple mountain ridgelines.

However, if your instincts tell you to honor the risk, then you'll find a way to fit Elecraft KX2 transceivers with KXAT2 automatic antenna tuners into your budget. Or, perhaps, there will be a mix of a few KX2 and several MFJ-9440 transceivers.

You now have all of the level 2 options for radios and repeaters. In the next part of the book, we'll discuss even more powerful HF transceivers for use in mobile and fixed command base stations along with a broad range of other supporting technologies.

Part 4 – Level 3 Command and Control

12

Strategic and Tactical Planning

Dear reader, if you're a newbie to two-way radio communications, you may presently feel as though you're pinned down at the bottom of an ever-growing technological dog pile. If so, this is expected, given that we've covered a lot of ground on the basics. These concepts were then applied in our discussions in the previous parts, Level 1 – Emergency System Radios and Level 2 – Field Radios and Repeaters.

In Level 3 – Operations and RFE Broadcasting, we're going to learn additional new concepts as we begin to crawl out from beneath this dog pile.

Level 3 of your three-tier community communications strategy includes the command and control for your entire communication system. This will include strategic alliances with other survival communities, municipalities, and the eventual creation of your own local Radio Free Earth broadcasting network. How is this so?

The radio basics that we've covered have provided you with the core concepts necessary for creating a solid foundation for all other aspects of your community communications strategy. With level 1 and level 2 you applied these core concepts in a tactical manner for various mission roles. With level 3, we will build upon all this with the tools crucial for strategic planning that you will need as a survival community leader.

Let's go back to that dog pile and consider what put you beneath that technological dog pile in the first place. What is it? It's the differences between tactical and strategic planning.

Strategic vs. Tactical Planning

For the purposes of our discussion, let's define the terms strategic and tactical and their use in the planning necessary to implement a successful community communications strategy:

- **Strategic Planning:** A strategic strategy is an overall campaign plan with a long-term focus that addresses all of the knowable elements of a complex operational and decision making process.

- **Tactical Planning:** Typically implemented by commanders in the field, these plans are short-term by nature. They are used to organize the actual means needed to gain a strategic objective.

  This difference is important. If you are only focused on the tactical issues and requirements without the context of the larger strategic purpose, you will smash into brick walls you never knew existed.

  An example of this is the number one reason many newbies give up even after they obtained their technical class HAM license and an inexpensive VHF/UHF two-band handy-talkie. They go to all that trouble only to shovel everything into the back of a dusty drawer along with their interest in amateur radio. Perhaps they'll eventually revisit amateur radio, but the sad truth is most that leave this way never do. Like the proverbial one-trick-ponies, they come and then they go.

This is a conundrum for amateur radio. The vast majority of those who give up this way are 30 or 40-something men and women in their prime adult years. These are the very people amateur radio organizations, such as the American Radio Relay League (ARRL), need to maintain a stable base of HAM members well into the future.

In February 2018 and for this reason, the ARRL asked the FCC to provide HAMs with a technical class license greater access to various bands and modes in the HF spectrum. They were asking for enough to give them, hopefully, a reason to stay engaged. This was a smart move by the ARRL.

With the current FCC band plan that is currently in place, technician class licensees can work the VHF and UHF amateur bands and frequencies above 30MHz. This license also includes limited CW (Morse code) use on the 80, 40 and 15 meter HF bands. Technicians can also work the 6m and 10m bands for the phone and digital modes. Overall, it's spotty though.

What does this mean for those holding technician class licenses but do not know Morse code? Not a lot of choice unless they upgrade their technician license to a general license for more access. If studying that much is discouraging, there are alternative options like the quad band Yaesu FT-8900R mobile radio. This transceiver is specifically designed for the restricted technician license band plan frequencies allocations in the 2m (VHF), 6m (VHF), 10m (HF) and 70cm (UHF) bands.

How many newbies know about the Yaesu FT-8900R? Darn few, they usually prefer to buy a much less expensive dual band VHF/UHF HT; their aims are predicable. They want to test the amateur radio waters while expending as little effort, money, and time as possible.

This is easier now than ever. Passing the technician license test now has many test preparation resources which are available on the Internet and through local clubs.

Once newbies get their technician licenses, call signs, and HTs, they can start trying to make local contacts. That's when the "bloom is off the rose" as the old saying goes because they quickly learn that amateur radio is not as easy as sitting in a reclining chair and talking to the world on a smartphone.

Rather, what they find out the hard way is that to make a contact with another local HAM, they usually have to step outside in order to reach a local repeater. This assumes that they live within the line-of-sight of one. As for those living out in sparsely populated rural areas, finding any repeater with an HT is often impossible.

Consequently, the practical value of a technician license plus an HT is next to nil for smartphone newbies. It doesn't take long for the license, the books, the HT, and whatever else to be penciled under the "cut your losses" column.

So how is this predicament relevant to this chapter, let alone this book? Going the technician license and HT route is a classic example of a naive, cheap-and-quick tactical plan gone wrong. It serves no useful strategic plan.

Now dear reader, stop and think for a moment about everything that you've learned up to this point. The strategic concepts presented in this book are intended for those in preparedness who view two-way radios as a necessary survival tool. This is why the level 1 and level 2 concepts (presented previously) do not require a general class license from the FCC. Everything can be done with a technician license or no license at all. The vast majority of folks in your community will be serious about survival – not amateur radio. They'll just want to press a button and talk, and for the most part that is how your community communications strategy will work with levels 1 and 2.

The goal with level 3 is to create a nexus of awareness for your community and alliance partners through the use of two-way radios and broadcasting. This means, that at level 3, the members of your community must be fully committed to making the system work. They must learn what they need to learn and study what they need to study. In other words they must show initiative. This will not work if you must threaten to send them to bed without their cookies and milk if they do not finish their homework. Therefore, this chapter will offer a different conversation along with congratulations as well.

As you read these words, we both know that you likely feel as though you're being crushed under a technological dog pile. Yet, you haven't given up and retreated to your smartphone enabled lazy chair. Rather, you're here; you're committed; and you have vision. In other words, you have what it takes to make the people work; so that the people can in turn make the technology work. So let's start this new leadership conversation and begin with the need to create a culture of communication within your survival community.

Culture of Communication

At the outset of the global tribulation, each of us will have the same three choices. We can be either part of the solution, a part of the problem, or a victim. If you and the members of your survival community are resolved to being part of the solution, then you must also be committed to working with local municipalities and other survival communities with whom you share common goals and mutual respect.

Therefore, being perceived as a problem is not a desirable outcome for any survival community. This raises an important question. What constitutes a viable survival community that is part of the solution?

Self-Defense

Viable survival communities during the global tribulation will typically seek peaceful existences. Most will be either egalitarian with the belief that all are equal or faith-based. They will be located in unincorporated county areas and outside nearby cities or towns. The majority of the memberships will be comprised of multi-generational families that are centered around healthy and productive young adults.

The optimal starting populations for these communities will be between 150 to 200 members. These communities will form alliances among themselves and with local municipalities. While health and welfare will be a long term requirement, the overarching motivation will be the need to collectively deal with predators and the other dregs of society. These include gangs, thieves, renegades, cannibals, warlords and other such post-apocalyptic threats.

Although Hollywood pumps out post-apocalyptic sci-fi films that present good people as easy pickings for predators, in a global tribulation world the good will be anything but weak. These communities will attract noble warriors with ample training, the skills and experience in hunting, sharpshooting, and sniping. They will know that the best defense is an offense, and they'll be the ones who will venture beyond your outer perimeter to hunt down and destroy predators long before they can get to your inner perimeter. Here is where your level 2 communications will play a vital role in assisting in reconnaissance, surveillance, and attack coordination.

There will be no due process or Miranda rights for predators. Most will die rather violently long before the report of a sniper's rifle can reach their lifeless bodies. Eventually, these predators will die off or feed upon themselves. That is when the principal need for survival communications will shift from self-defense to self-reliance.

The need for security and situational awareness will be an essential survival requirement at the outset of a global tribulation. This is not a new concept by any stretch of the imagination. Rather, it dates back to our early hunter-gatherer existence where food was hunted, fished, or gathered.

A History of Situational Awareness

During the early part of human existence, there was a practical division of labor between the sexes. Men formed hunting and fishing parties. Women would mind the children as they gathered foods such as fruits and berries.

The men during the hunt understood the need for stealth so they spoke few words, preferring instead hand and face gestures. On the other hand, women spoke to each other continuously as they went about their daily business.

This is a clear example of situational awareness. The women were not gossiping. Rather, it was their culture of communication, and for the women it proved to be the best way to keep those not hunting with the men alive. The strategy was simple. If one woman unexpectedly stopped talking, the others knew there was a problem and took action.

Coming forward to these present times, if you visit a HAM radio club meeting, the first thing you'll notice is that few women are in attendance and most are there because their husbands are. They are always treated well and made to feel welcome.

However, the women and girls who share the same level of interest as the men are treated with the greatest respect, and the fellas are always happy to help out. It is a heartwarming thing to see this. Even though chivalry may be a thing of the past in other areas of modern life, it is alive and well in amateur radio. So who is doing most of the yakking on amateur radio these days? It's still the guys and by a good margin, but what about the future when amateur radio is no longer a male-dominated hobby for the technically inclined?

If your community is to survive a lengthy global tribulation with its noble virtues intact, your first responsibility as a leader is not to write checks. It is to create a culture of communication that is not just about men and women. It must be about entire families.

This is why you must always seek ways to involve and engage women and children in as many communication roles as possible within your community. Women for example have long enjoyed working together in activities like quilting circles. During a global tribulation, quilting will be an essential survival skill.

The same women who make quilts for their families, friends, veterans, and so forth are already organized. All that is needed is to substitute quilting materials and tools with wire antenna fabrication tools, supplies, and testing equipment. Then, once they've been trained, they'll be able to fabricate wire antennas independently; they'll also manage themselves; and along the way, they'll come up with new ideas that will delightfully surprise.

The bottom line is that preference must always be given to any female in your community who is ready and willing to step up to a communications role on an ongoing and committed basis. It is not that we do this today; in the future it will be the combination of this and other survival communication principles that will greatly enhance your ability as a leader to ensure that your community can survive and thrive.

Survival Communication Principles

The difference between a local or regional catastrophe and global tribulation is that with a local or regional catastrophe, those living in unaffected areas can watch the catastrophe unfold on television as they donate money to the victims using a credit card.

In a global tribulation the usefulness of these infrastructure dependent resources will disappear like tears in the rain. When this happens, most of the –ologies and –isms that presently define our lives will become irrelevant or unhelpful.

Your community during this time will likely be outnumbered by adversaries who will be able to sustain higher combat loses than your survival community. Therefore, regardless of your faith or beliefs, the following survival communication principles you should keep in mind are:

- Fast Decision Making
- Internal Cohesion
- Human Capital
- Signals Intelligence
- Self-Dependency

When incorporated into your community communications strategy, these principals will give you a powerful communications edge. So, let's take a closer look at each of them.

Fast Decision Making

A classic example of the effects of slow decision making is D-Day (June 6, 1944) when the Allied forces invaded northern France at Normandy. Well in advance of the invasion, Nazi General Erwin Rommel made extensive preparations and plans for destroying the invasion on the beaches. Yet, there was no German counterattack.

Had Adolf Hitler given the command to launch Rommel's plan, the Wehrmacht could have defeated the Allies on the beaches. Hitler's strict orders not to be awakened that day were a windfall for the Allies. Even if Hitler had been awakened, he probably would not have ordered a counterattack at Normandy because he was convinced that the actual invasion would come at France's Pas de Calais region. The rest is history.

The point here is that despots like Adolf Hitler who hold absolute power all have the same failing. They need their armed forces, but not at the expense of a coup d'état conducted by a local commander with autonomous authority. Therefore, they create rigid hierarchies that deny local commanders the autonomy needed to make decisions on the ground.

The lesson to be learned from this example is that communication alone does not bring victory. Hitler had the benefit of a robust communication system and squandered it with his paranoia and his incompetent micromanagement of the military. In fact, after D-Day the Allies could have assassinated Hitler but chose not to do so. His rash incompetence as a military leader had turned him into the most valuable of war-winning assets for the Allies.

Communities ruled the same way during the global tribulation will be just as vulnerable to defeat as the Nazis were on D-Day. Therefore, your local commanders must have the autonomy to make boots-on-the ground decisions.

This requires a flatter and less hierarchical organization. Commanders need to communicate with other local commanders and community leadership so that a real-time collaborative effort to combat hostiles and predators is made long before they reach the community's outer perimeter.

To help you with this concept, there is a useful bit of knowledge for developing your community's communication strategy. It is called the power distance index (PDI). Created by Dutch social psychologist, Geert Hofstede, the PDI refers to the way in which power is distributed within a community. How much is held by the leadership and how this distribution of power is perceived by the less powerful.

When the power is distributed unequally with a tall and rigid leadership structure (such as seen with Nazi Germany) the PDI would be high. This would then indicate that the top leadership would have an inflated sense of authority and be slow to make timely decisions. The leadership at the lower levels would then be highly cautious to show initiative, and this would impair their ability to assess and initiate things independently.

Conversely, in a less hierarchical culture, the PDI would be low which would indicate that the leadership at all levels would function in a more real-time collaborative mode and consequently make faster decisions. The huge benefit here is the initiative.

A low PDI offers a smaller sense of leadership and initiative is encouraged as a foundation for victory – not a potential coup d'état. While combat opportunities to seize the initiative may be fleeting, if seized quickly by a local commander, they can carry the day.

For these reasons, designing a leadership strategy that has a low PDI score should be your goal. Then you amplify the winning power of that with a community communications strategy that supports your local commanders in the field. Your field commanders will then have the ability and the autonomy to seize the initiative whenever the opportunities present themselves. Then the sage advice of Sun Tzu will apply — "Opportunities multiply as they are seized."

Internal Cohesion

Cohesion is what makes a survival community viable. That happens when everyone eats, sleeps and fights as a team. Even the children must have a job and they must do it. For example, a ten-year old with a bolt-action .22 rife can become a very deadly warrior. As far as the younger children, they can work with their grandparents to retrieve, load and distribute rifle magazines.

Conversely, survival communities with a high PDI score will be weighed down by special privilege, elitism, sexism, and other divisive practices; all of which will result in a serious loss of cohesion.

Simply put, the men of high PDI communities will be prone to using their inflated sense of authority to enslave and sexually abuse women and children as they please. The last thing they would ever do is give a weapon or a two-way radio to any woman or child; lest they sacrifice some or all of their ability to abuse-at-will.

This is a critical point. According to conservative estimates, one in four children today has been sexually abused by an authority figure. This is a really terrible plight upon our nation.

The corrosive effect of this reality will be even worse because abused women and children will not fight as part of a cohesive internal community. Rather, their will to survive will likely compel them to put their survival above that of the community.

Herein, will be an Achilles' heel that can be exploited. Take in a few of these abused women or children, then treat them well and give them a home. They will happily give you the inside intelligence you can then use to exploit the weaknesses of a predatory adversary.

However, when your community communications strategy empowers women and children with clearly defined communications roles, internal cohesion becomes possible in a very significant and powerful way.

The bottom line with predators like Adolf Hitler is that they do not eliminate divisive loyalties. They would rather embrace them and the uncertainty they create. Why? So they can suppress any challenges to their power which results in an existence of leverage, force, fear, and an uninspiring future.

A survival community leader on the other hand who actively pursues a solid and internal cohesion strategy will foster a community-driven combat morale like the one offered by Alexandre Dumas in *The Three Musketeers* – "All for one and one for all, united we stand divided we fall." This is how a bright existence of love and commitment can be nurtured and also prized by all of those who will endure and those who will sadly fall, while in service to a viable survival community.

Human Capital

Several assumptions are necessary for survival during the first part of a global tribulation. The membership of your community will be outnumbered by the bands of predators, and the leaders of these bands will see their members as expendable and will deploy them as such. Your community, which is outnumbered, can ill afford to lose its people in such a reckless and wasteful way. Therefore, they need the best available equipment and training that can be provided.

Equipment

Before China joined the World Trade Organization (WTO) in the last century, Sears stores had a practice of categorizing products with a "good, better or best" designation. The products in all three categories were durable, and the principal differences were the number of available functions, features, and cosmetics.

Today, we live in a "throw-away" world and examples of this are the bargain-priced, "throwaway" Chinese brand radios. The whole proposition is flipped upside down in this case where these manufacturers prefer the number of available functions, and features, and cosmetics over product durability. Building real durability into a product is expensive.

This is where technically-challenged consumers tend to make false assumptions. If a manufacturer offers a 2-year warranty, it is often assumed that the product will last for two years or more before it fails. That may work for plastic storage bowls and coffee grinders but not for electronics.

This is why the mean time between failures (MTBF) is what you need to keep in mind when evaluating any consumer or amateur transceiver or radio. Engineers and enthusiasts use the MTBF rating provided by manufacturers to predict the amount of time that will lapse between inherent failures of a product during normal operation.

Therefore, the operative term here is not warranty. It is how soon the product will typically fail in the course of normal operations. Generally, electronics will most likely fail right out of the box or within the first thirty days, if it is going to fail, it will be early on.

Therefore, if you turn on a cheap "throw-away" transceiver a few times to test it out for an hour or so you are fooling yourself. Even worse yet, you then pack it away in a survival kit for two years; all you will have then is a dust magnet warranty for a product that will fail and most likely when you need it most.

Sure, There are many who like these cheap "throw-away" models, including Amateurs, but how are you going to make a warranty claim in the midst of a global tribulation? That's the issue when you are buying equipment for preparedness.

The point is that no matter how sexy a radio or transceiver looks with its oodles of bells and whistles, do not be suckered in. Always begin with durability and simplicity and get the features needed to perform the mission with as few but necessary bells and whistles as possible. If you can perform a mission with an affordable MIL-STD single band transceiver, that's your baseline. That's what you look for. Start with good (as in durable). If you need to step up to better or best, then, it is because you need those features and functions for the mission. It is not because they are sexy and will give you bragging rights.

Training

There is an old saying in amateur radio. "Some things must be caught – not taught." A good example is learning Morse code. You must first learn how to hear it and then learn how to key it. Therefore you must train two of your five senses through repetition.

To illustrate the concept, imagine that you are standing in line at a local bank. A few young and very nervous fellows rush into the bank to rob everyone at gunpoint. One of the robbers suddenly pulls you out of line and then puts a pistol to your head.

As the sweat beads on your forehead, out of the corner of your eye you can see through the window and across the street. There, on the roof of the building is a SWAT team marksman with an expensive high-powered sniper rifle. Meanwhile, the thug holding a pistol to your head is starting to get shaky and you're beginning to doubt that you'll live to see the end of the day.

Would you want that SWAT sniper across the street to get on a radio and say to the incident commander, "My uncle got me this job and I just started today. I could use a little help. Could someone tell me which way the pointy end of the bullet goes?"

Or, would you want to know that that sniper has been practicing at the range for years, for situations precisely like the one you're in? And that sniper has a dead aim on the thug holding a pistol to your head?

What do we mean by dead aim? The sniper has the thug's apricot lined up in the cross hairs of the rifle's scope. Snipers use the term "apricot" to describe the medulla oblongata at the base of the brain stem. Among the many things it does, it helps transfer neural messages from the brain to the spinal cord. The sniper knows that the perpetrator dies instantly and your head is not blown off once his bullet tears through the apricot.

This is the point of training for survival missions during the coming global tribulation. Reading a book and putting it and a radio in the drawer and then forgetting about it until you need it is a good way to get yourself or someone else killed. Rather, it is imperative that you train your people how to use whatever equipment they're assigned for particular missions.

What you're doing is creating muscle motor memory through repetition. The sniper does not have to think about breathing, or how to hold the weapon, and so forth. That is what is done in training. You also need to create new neural networks of your own. The human brain is an amazing organ that can program itself. The more you contemplate a potential situation or circumstance makes you better able to deal with it should it happen. It is through contemplation and repetitive training that your brain builds new neural networks to address this situation.

Another example of this is what happened during the 2004 Indian Ocean earthquake that claimed over 270,000 lives. What was learned after that tragic event? The people, who did most of the dying in the Sumatra region, were the modern people. Folks with new cars, good jobs, cash in the bank, and so forth. They were completely unprepared for what happened. When they saw the water receding from the coast, they just watched in stunned amazement. When it came back in, many died badly along with their pets.

However, native tribes and wild animals living on nearby islands did not die. The moment the first trembles were felt these folks all did the same thing. They started running uphill as fast as they could screaming, "Feet don't fail me now." Why? They grew up in a culture of communication where the tribal elders taught them what to do in different circumstances.

Did they talk like us and use clever technical terms? No. They would say things like, "When the big mother turtle of the world trembles, stop what you're doing and climb the shell." It doesn't matter how corny that sounds; if it works, it works. This communication of culture allowed them to build the neural networks over time. Consequently, with the first tremble, the tribal members did not have to think about what to do. They acted; thanks to the neural nets created in their minds. Then, as the wild animals ran uphill, so did they – and while smart modern people were drowning like rats, the natives survived. How smart is that?

Now imagine that you have a team operating in the field beyond your outer perimeter and they are ambushed and need assistance. Not only should the field radio operator assigned to the team know how to transmit a distress message, each team member should also know how to do the same.

They should be trained so proficiently that they're not going to need to thumb through a manual while bullets are flying over their heads. Each team member must be able to pick up a radio and know exactly what to do in time to save lives.

Knowing when you're walking into a trap is of equal importance.

Signals Intelligence Gathering

A common mistake of many newbie HAMs is making the assumption, that if they do not hear anyone talking on a given frequency, it is not in use. However, once they announce they are looking for a contact, another HAM will speak up and the contact is made.

This may be surprising for the newbie, but the truth is that many people or perhaps hundreds of people are usually monitoring that frequency at any given time. The critical question for you is not why are they silently monitoring that frequency? The critical question for you is what are you monitoring?

The Imperial Japanese navy during WWII was dealt a massive blow at the Battle of Midway in 1942 when it lost four major aircraft carriers. Their plans to rule the Pacific Ocean after that were eviscerated.

Nazi U-boat wolfpacks in the Atlantic were sinking massive amounts of Allied shipping, and the Nazis nearly brought England to its knees. However, the tide turned in mid-1942 in favor of the Allies as new sub-hunting technologies came to bear. Many brave sailors gave their lives in both theaters of combat, but the deciding factor in both cases was signals intelligence.

Before the Battle of Midway, the US Navy broke the Japanese naval codes and knew their battle plan and where to engage them. This advantage enabled two American carriers to sink four Japanese carriers with the loss of only one American carrier.

The British in a similar vein had managed to break the German enigma code and were able to reroute convoys around the German wolf packs. Had the Allies been unable to break those codes, we could be living in a very different world today.

The Germans and the Japanese were confident that their codes had not been broken, and this arrogance led them into traps of their own making. This is why signals intelligence gathering will be essential to the success of your community communication strategy. As author Frank Herbert advises so elegantly in his book Dune, "Knowing that a trap exists is the first step in avoiding it."

Information is power. Not everyone talks. Few do. Most listen. If you're too lazy to listen because you think it is a waste of time, then you're lazy enough to die. It is that simple.

Self-Dependency

Since the early days of amateur radio, the most important virtue of this technology was infrastructure-independence. There were no middlemen, and no apps that require a Wi-Fi connection to the Internet, etc.

Today, amateur radio is transitioning to the digital age, and while this offers many advantages, it comes at the terrible price of infrastructure-dependence. When the infrastructure fails, your two-way digital radios will also. Then, all the money and time you spent on infrastructure-dependent technologies will circle the drain along with your survival advantage. Ask a local sheriff what will happen to their communications when the infrastructure fails? Do not be surprised if the answer is a shrug of the shoulders and "I don't know."

Therefore, to ensure long-term success, your community must be as self-dependent as possible and in as many different areas as possible. Therefore, you should seek out simpler and time-proven designs. They may not be as sexy as the new stuff, but when it comes to survival, they will work. Your community needs to make as many components as possible, like wire antennas, on its own.

The point here is that you should organize your communications to be as self-dependent as possible. The simple things you can do are better than the complex things you cannot do.

Summary

In the two previous parts of this book, you were presented with a tactical introduction of the types of communications gear you'll need for level 1 and level 2. However, level 3 from a leadership perspective is very different because you're dealing with human engineering more than radio engineering.

If you and your technical team create a communication system that is complex, unwieldy, and prone to failure, the membership of your community may become woefully short of patience and confidence. This would especially be the case if you and your technical team begin to look as though you're floundering.

Therefore, the foundation for level 3 must be built upon confidence, but if the truth be known, you can expect some floundering at first. Sure, it will look embarrassing, but if the members of your community are confident that you are on the right track, they'll track along with you.

To begin building this confidence, you must demonstrate a competent and consistent approach to assessing needs and options. What does that mean?

As the leader of a survival community you do not need technology first.

Rather…

You need a visionary process that gets you to the right technology!

13

Assessing Needs and Options

The previous chapter ended on the need for a simple visionary process that will build confidence in your ability to create and manage an effective community communications strategy. That being:

>As the leader of a survival community you do not need technology.

>Rather…

>You need a visionary process that gets you to the right technology!

This visionary process statement was repeated because it serves as a problem-solving framework. Not only for you as a leader, but for all of your community leadership as well.

Always remember that trying to impress people with what you think you know about two-way radio technology could lead others to a nagging concern about your focus as a leader. This would be a most corrosive outcome indeed. Therefore:

- Asking intelligent questions is your job.
- Responding with intelligent answers is the job of your technical team.
- The end goal of this process is a totally informed decision.

There are no guarantees in life. Everyone knows that and that even a totally informed decision can fail. As the old saying goes "some days you eat the bear and some days the bear eats you." What they'll need to see is competence over time and the best way to do that is for everyone to see how you intelligently assess your needs so as to find the best available options for any given communication mission.

Assessing Needs

Before you can assess the best of all available options for a communications mission, you must first assess the mission needs. This will prepare you for the second step in the process, in which you will determine which region of the radio frequency spectrum to use.

This chapter will review concepts previously introduced about the various aspects of the radio frequency spectrums. But for now, what is most important is that you build confidence in others with a competent and consistent approach.

What will this confidence feel like for those observe your approach?

To illustrate the answer, imagine you are leading a field team that is operating far beyond the outer perimeter of your survival community.

When you press the press to talk (PTT) switch on your microphone, and announce your message, you may feel as though you're dangling out on a shaky limb. Then, you hear the command base station's reply. This is when you know that you are a part of something bigger than yourself and this will empower you.

Next, imagine you standing in your command base station and hearing many such contacts being made each day. As you do, you will know you are the nexus of that something bigger because this is where you can best serve your community.

To get to that point, hearing what others have to say is where you begin. Not only must you gather the right information, you must do it in a consistent manner that builds confidence, which means that people need to feel that you understand their needs and concerns.

While there are libraries of useful books on this topic, there are two simple things you can do and teach that will give you the most important confidence-building tools. These are the "Five W's and How" and Power Listening Skills.

Not only will you use and demonstrate these confidence-building tools, you will demonstrate their effectiveness and impress them upon everyone in your community who will choose to participate as correspondents in your local Radio Free Earth network.

Five W's and How

Professional journalists use the Five W's and How to gathering the facts needed to assess a situation and then to create a complete news account. Those who pursue a college degree in mass communications learn the Five W's and How in their first year. Here is what they learn:

- **Who** was involved?
- **What** happened?
- **When** did it take place?
- **Where** did it take place?

- **Why** did that happen?

- **How** did it happen?

Each of these six questions must be answered factually and in detail and "yes or no" answers are wholly unacceptable. Likewise, talking to hear yourself talk is equally unacceptable because when it comes to asking questions, less is more.

Keep your questions simple, polite and professional. Then use five power listening skills to convey your ability to arrive at conclusions through a thoughtful and well-considered process.

The Five Power Listening Skills

In 1968 the Beatles traveled to Northern India to study Transcendental Meditation (TM) under Maharishi Mahesh Yogi. This set in motion an amazing pilgrimage of spiritually-challenged Westerners flying to India for wisdom.

The cartoonists of the day had fun with it and in the newspapers we saw cartoons depicting clueless Americans climbing tall mountains in northern India to humbly stand before a guru in search of wisdom. Inevitably, the gurus all listened patiently, then having reflected on what was asked, would respond with a simple statement or another question.

This resulted in making the gurus look incredibly wise even though they spoke very few words. While some can debate whether these gurus are truly wise or not, what they did to make themselves appear wise was what I call the five power listening skills. They are:

- **No Clipping:** The most insulting thing you can do is to interrupt with a response before a person can complete what they are saying. This sends the message that you have no respect for the person you are listening to. They in return will have no respect for what you have say as well. They'll just be busy thinking how rude you are.

- **No Anticipating:** If you are thinking about how to respond to someone as they are speaking, you'll miss important details. Consequently, when you answer you will likely make false assumptions based on a part of what you're heard and you will appear arrogant and uncaring. This is a real audience interest killer.

- **Listen Thoughtfully:** When you answer questions with angry or negative emotions, you will stifle or silence those who have something honest to say. Conversely, those who seek gain through manipulation will know how to make use of your anger, to their benefit. However, by listening thoughtfully to someone with an attentive and calm demeanor as you take note of what they're saying, you will encourage them to be forthcoming.

- **Pausing for Focus:** Never forget that the perception of wisdom always begins with silence. You can only listen thoughtfully when you are silent. Once a speaker finishes

what they have to say, continue to maintain direct eye contact, but remain silent for a few seconds before you answer. As a rule, count to five and then respond. This will give others the impression that your answer is well-considered and wise.

- **Ask Confirming Questions:** Of all five listening skills, this is the most powerful of all. This is because instead of demonstrating what you are ignorant of, you engage the speaker with questions and thereby find clarity. For example, such a question can begin with, "It is important to me that I understand you properly..."

Use these power listening skills and you will be seen as cutting through all verbiage, straight to the heart of the issue. It is a less is more proposition. By using fewer words, you will be seen as being wiser and you will also build an expectation of a wise decision.

In most cases there is usually time to make a decision, whether at the close of the dialogue or later, after taking the matter under advisement. (This is just a tip.) If possible, wait to see which options materialize at the eleventh hour. Invariably, this is when the best options tend to appear.

However, before you assess the best options, you must complete your needs assessment. These findings will help to define right options.

Assessing Options

In the radio basics part of this book, we covered the defining aspects of spectrum regions, services, modes and frequencies. In the table below, Spectrum Regions and Services, these defining aspects are presented in the context of this leadership assessment process.

As you read through this table, take note of the channelized and amateur bands. You'll see the bands expressed as 40m or 20m for example. With a 20m band, a full wave antenna will also be 20m (66 feet) long. In the frequency range column of the table, you will also see the frequency range of the band, expressed in Kilohertz (kHz) and Megahertz (MHz).

As you read through table, take notice of the inverse relationship between bands and their associated frequencies. That being:

- As wavelength increases, frequencies decrease.
- As wavelength decreases, frequencies increase.

Also take notice of how AM and FM broadcasting are situated relative to the other services and bands you'll use with your own community communications strategy.

Spectrum Regions and Services			
Channelized	**Amateur**	**Broadcast**	**Frequency Range**
Medium Frequency (MF) - 300-3000 Khz			
		AM	535 – 1605 kHz
	160m		1.8 – 2.0 MHz
High Frequency (HF) – 3 to 30 MHz			
	80m		3.5 – 4.0 MHz
	60m		5 MHz channels
	40m		7.0 – 7.3 MHz
	30m		10.1 – 10.15 MHz
	20m		14.0-14.35 MHz
	17m		18.068 – 18.168 MHz
	15m		21.0 – 21.45 MHz
	12m		24.89-24.99 MHz
CB / SSB CB (11m)			26.9650 MHz – 27.4050 MHz
	10m		28 – 29.7 MHz
Very High Frequency (VHF) – 30 to 300 MHz			
	6m		50– 54 MHz
		FM	88 – 108 MHz
	2m		144 – 148 MHz
	1.25m		222 – 225 MHz
Ultra High Frequency (UHF) – 300 to 3000 MHz			
	70cm		420 – 450 MHz

Spectrum Regions and Services			
Channelized	**Amateur**	**Broadcast**	**Frequency Range**
FRS/GMRS (65cm)			462.5625 – 467.7250 MHz

The radio spectrum regions not included in the table above reside immediately below the MF and above UHF spectrum regions. They are worth mentioning as they may come up in conversation.

- **Below Medium Frequency (MF):**   There are two lower spectrums, very low frequency (VLF) and low frequency (LF). VLF is used for specialty missions such as radio navigation, time signals, and secure military communications. It is best known for its use with submarines which can send and receive in this band at a depth of 40 meters (120 ft.)   LF is used for aircraft beacons, navigation and weather systems.

- **Above Ultra High Frequency (UHF):** There are two higher spectrums, the Super high frequency (SHF) and Extremely High Frequency (EHF). SHF frequencies are referred to as microwave and this spectrum region is used for radar, satellite communication, microwave relay links, and wireless devices. EHF frequencies are referred to as Millimeter waves and are typically used for radio astronomy and atmospheric sensing.

  Above the EHF is where the optical spectrum resides which starts with infrared light and above that are visible light, ultraviolet and so forth.

The radio spectrum regions you'll use with your community communications strategy will all offer one or more phone modes. These will determine the range and quality of your signals.

Phone Modes

During the global tribulation, 70% or more of your radio traffic will be via a phone (voice) mode, across a broad range of spectrum regions, services and modes. The remaining 30% will be other modes such as CW (Continuous Wave – Morse code), RTTY, Data and so forth.

For these reasons, the first step to finding the right options for a mission with a phone mode requirement is to find the right spectrum frequency and phone mode. The phone modes are:

- **Amplitude Modulation (AM Mode):** This is the earliest phone mode, and it is the least efficient way to transmit a phone (voice) signal. AM uses a carrier wave with two sidebands. The carrier wave is in the middle. Above it is the Upper Sideband (USB) and below it is the lower sideband (LSB).

- **Single Sideband (SSB Mode):** This mode is an AM variant and it is the most efficient way to transmit a phone signal. With SSB one sideband and the carrier are suppressed, leaving the remaining sideband active. That being the LSB or USB. As a general rule, each spectrum region band will offer either one as proscribed by law or convention.

In the case where the sideband is determined by convention, Amateur Radio Associations and Manufacturers provide an industry guideline whereby they publish the assignment of a sideband as default. The convention is that LSB is used with frequencies from 0.03 MHz to 9.5 MHz and USB is used with frequencies from 9.5 MHz to 60 MHz.

● **Frequency Modulation (FM Mode):** In terms of transmit efficiency the FM mode is better than AM and less efficient than SSB. FM is the most popular phone mode for the 2m, and 70cm bands (VHF and UHF) and is supported exclusively by amateur radio association local repeaters all across the country.

Is it possible for transceivers designed for FM phone mode on the 2m, and 70cm bands to also receive and transmit in other phone modes such as AM or SSB as well? Yes, but only if the transceiver is designed to operate this way, in which case it is called "all mode, all band."

All mode and all band transceivers are the proverbial kitchen sinks of amateur radio and while it may be appealing to have just one transceiver that can do everything, always keep in mind an old saying, "Jack of all trades, master of none." Then there is cost, as all mode, all band transceivers will cost 3X to 4X as much as a dual band transceiver like the Kenwood TM-71A.

However, if this kitchen sink design strategy appeals to you, then you may wish to employ a less expensive way to evaluate the use of an all mode, all band mobile transceiver with your community communication strategy. Here, the used marketplace is the best place to start.

Among knowledgeable HAMs, the most desirable "all mode, all band" used mobile transceiver is the ICOM IC-706. Discontinued in 2009, the MIL-STD 706 was built like a tank and became one of the most popular transceivers in amateur radio history. The last version was the IC-706MKIIG which is still considered the best of the line.

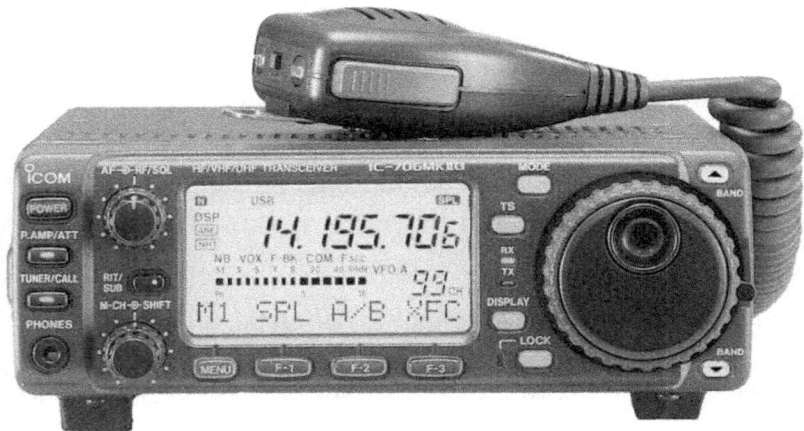

To fully appreciate the trade-offs between a dual band FM mobile transceiver like the Kenwood TM-V71A and the more complex all mode, all band ICOM IC-706MKIIG, let's not focus on price. Rather, let's use power as our measure as shown in the tables below.

Kenwood TM-V71A – 144/430 Mobile Transceiver			
Input Power (VDC)	**Current Drain Amps (A) Milliamps (mA)**		**Transmit (Watts)**
VDC Input Range	**Receive Drain**	**Max Transmit Drain**	**Maximum Output**
13.8 V DC ±15 %	1.2 A	13 A	50 Watts

ICOM IC-706MKIIG – HF/VHF/UHF All Mode Transceiver				
Input Power (VDC)	**Current Drain Amps (A) Milliamps (mA)**		**Transmit (Watts)**	**Input Power (VDC)**
VDC Input Range	**Receive Drain**	**Max Transmit Drain**	**Maximum Output**	**VDC Input Range**
13.8 V DC ±15 %	1.8 A	20 A	HF/50 MHz – 100W VHF (2m) – 50W UHF (70cm) – 20W	HF/50 MHz – 40W VHF (2m) – 20W UHF (70cm) – 8W

How do the numbers compare. In standby monitoring mode, the IC-706MKIIG requires 0.6 more amps than the TM-V71A. However, of greater importance is the maximum transmit drain. The IC-706MKIIG requires 8.0 more amps than the TM-V71A. What do you get for all this heavy power consumption? If you're working the UHF (70m) band, the maximum transmit power of the IC-706MKIIG is comparable to a Handy-talkie (HT) and an expensive one at that.

Another trade-off is complexity. All mode, all band transceivers are complex by design and are much more difficult to repair than simpler transceivers. For this reason, always use the KISS (keep it simple stupid) strategy for selecting a suitable transceiver for any communication mission. Here is where the characteristics of the three phone modes will be essential in determining your best options for 70% of your communication mission needs.

Phone Mode Characteristics

Your phone (voice) communications traffic will employ an AM, SSB or FM mode in one or more roles. Therefore, in terms of survival communications missions, it is important to understand the mode attributes: Sound quality, propagation and signals intelligence.

Modes and Sound Quality

If we use the sound quality of a modern smartphone as an example benchmark, how does it compare with the AM, SSB and FM phone modes used by the transceivers discussed in this book compare relative to sound quality?

- **FM:** With proper reception, the sound quality of the FM mode is very close to that of a smartphone or a broadcast FM radio station.

- **AM:** With strong reception, the general sound quality of the AM mode is comparable to that of an old dial-up telephone or a broadcast AM radio station.

- **SSB:** Of the three phone modes, SSB has the smallest bandwidth which makes it the most efficient for transmitting signals over long distances. However, newbies often complain that SSB mode has a hollow sounding voice. This complaint is often rectified with the use of a clarifier which fine tunes the signal for better reception.

The key sound quality differences between these three modes are bandwidth and modulation.

Bandwidth and Sound Quality

To illustrate the concept of radio bandwidth, let's use three different pickup trucks. One is a small quarter-ton, another a conventional half-ton, and then a larger one-ton truck. We'll use these trucks to haul aggregate to spread on the dirt road leading to your country cabin.

Aggregate is aggregate no matter how much of it you haul, so the issue is the load capacity of the truck. A small quarter-ton is rated to carry one fourth as much as a one-ton truck which means you'll have to make four times as many trips from the quarry to your home. Does this mean the one-ton truck is preferable? In this case yes, assuming you have enough gas, because ten gallons of gasoline will go a lot further in a quarter-ton truck than a one-ton truck.

So here are the trade-offs. How much aggregate do you really need to spread on your drive? This will determine how many trips you need to get all of the aggregate for the driveway, vs. how far you can go on 10 gallons of gas.

If we apply this to radio phone modes in terms of bandwidth, one could say that SSB mode is like a quarter-ton truck, AM is like a half-ton truck and FM is like a one-ton truck. After all, aggregate is aggregate and the same holds true with sound quality. How much do you need?

This is why the SSB mode (the smallest truck in this example) is the most efficient phone mode for transmitting signals. This is because with SSB, distance is more important than sound quality. To illustrate the point, let's begin with broadcast signals such as those we hear with broadcast AM and FM stations. By law:

- AM stations broadcast in America transmit in the MF spectrum region from 535 – 1605 kHz and the interval between each AM frequency is 10 kHz.

- FM stations broadcast in America transmit in the VHF spectrum region from 88 – 108 MHz and the interval between each FM frequency is 15 or 20 kHz.

We can use frequency intervals to help shape a mental image of the inverse relationship between bandwidth and sound quality. You could imagine frequency intervals like a truck's cargo bed. Greater sound quality requires larger bandwidth (as in a larger cargo bed), but at the expense of distance. This is not about compromises. It is about knowing what you really need.

This is why when driving far out in the country on our way to a country cabin, that we hear a lot of distant AM stations and perhaps a few FM stations, or none at all. What do we hear on AM? Talk radio, news, ethnic programming and so forth. If you like classical music, bring a CD or wait until you're close to a large city. The wider bandwidth of FM is required to transmit authentic-sounding music. That bandwidth is simply not there with AM.

Granted, the actual frequency intervals for other services such as amateur radio vary, but the same general concept applies. Therefore, in terms of amateur radio, how does the frequency interval of the SSB mode compare with that of the AM mode?

SSB Mode vs. AM Mode

The frequency interval for AM frequency is 10 kHz and this is comprises three parts, the carrier wave, an upper sideband (USB) and a lower sideband (LSB). The two sidebands are used to carry the signal intelligence, whether that is voice or music makes no difference. The carrier wave is then used to transport the signal intelligence on the two sidebands, but it also does something else.

The carrier wave is used to transmit the sound of silence. It is why we do not hear static and hiss between spoken words and breaks in the music as it is typically much stronger than the background noise and simply overpowers it.

Now let's apply this to SSB mode, a variant of the AM mode. The bandwidth of the SSB mode is approximately one third that of AM. As was discussed previously, with SSB, the transmitter removes the carrier wave and either the upper or the lower sideband. The intelligence is then concentrated on the remaining side band.

Unfortunately, by removing the carrier wave, the SSB mode also removes the carrier wave sound of silence and so the transmitter does not transmit anything when the operator is not speaking. Consequently, this is when atmospheric noise, local and remote electrical noise from devices and thunderstorms for example create interference. You can also pick up broadband noises called RF hash from computers and other electronics such as birdies or chirps.

Another issue is tuning. Due to the wider bandwidths of the AM and FM modes, once you've tuned in a station tuned in, there you are. However with SSB, since there is no carrier transmitted to lock the signal and prevent signal drift, a way to manually tune and synchronize the carrier inserted by the receiver is required. This is commonly called a clarifier control and allows the operator to fine tune the signal as needed for drift.

Yet, SSB is the number one phone mode of choice for Amateurs on the HF bands. Why is that? When they want to listen to opera with authentic, ear-pleasing sound quality, they listen to an FM broadcasting station. When they want to contact other HAMs around the world, all they need is enough sound quality to intelligibly copy the signal. In other words, transmit distance outweighs sound quality. Besides, SSB is easier to copy under many unfavorable conditions than the other two phone modes. So how does FM stack up against AM and SSB?

FM vs. AM and SSB

Of the three phone modes, FM has the widest bandwidth, which means that it requires the widest frequency intervals. (Remember, it is the one-ton truck in our example.) Of the four spectrum regions, MF, HF, VHF and UHF, FM is the dominant phone mode for VHF and UHF.

This is partly due to an outgrowth of amateur radio's early years, when HF was the only bandplan. At that time, the bandplan was designed around the more limited bandwidth of the AM mode. With the subsequent addition of the VHF and UHF spectrum regions decades later, the bandplan was expanded larger intervals so as to allow for greater bandwidth. This is why the FM mode can be used on the 10m HF band around 29 MHz, but nowhere else in the HF spectrum region.

Still the same, FM is presently the mode most new HAMs use to get started with via radio club repeaters on the VHF and UHF spectrum regions.

Another reason new HAMs like FM is the sound quality. For voice communication, the FM mode is exceptional as there is generally no noise or fading. Or in other words, the FM mode has an excellent signal-to-noise ratio (SNR)

HAMs use the SNR to compare this relative level of difference between a desired signal and the level of background noise mixing with it. Without a doubt, the SNR of the FM mode is significantly better but when it comes to signal propagation, FM has short distance legs so to speak.

Signal Propagation

Propagation is a term used by HAMs to describe the principles of electromagnetic radiation as they apply to two-way radio communications. In short, how does your signal get from you to whoever is receiving it and how far away from you can that person be?

Keep in mind; signal propagation is a complex topic with a wide number of granular variables such as the frequency used, transmitter power, antenna design, solar activity, weather, and so forth. However, for assessing the best options the three things you need to know are the three principal types of signal propagation: groundwave, skywave and line-of-sight.

DX Communication Propagation

With low frequency signals such as amateur MF and HF, long distance (DX) communication ranges are possible because of groundwave and skywave effects. On the other hand, higher frequency signals such as VHF and UHF only propagate well with a line-of-sight effect.

To illustrate this difference, let's use a home entertainment system. The sound waves from the system fill the house. You can walk around the house and still hear the sound in another room. In a rather simplistic way, we can imagine this as groundwaves and skywaves. They have the ability to convey the signal intelligence beyond the line of sight.

However, if you are using an infrared remote to control your sound system, you must always aim it at the infrared control sensor on the bezel of your television set. You can step out of the room, and even if you are aiming the remote in the direction of your television, the infrared signal of the remote will not penetrate the wall and therefore will not work.

Ergo, what this example shows is that while you can hear the music in the next room, you cannot control the television unless you have an unobstructed, line-of-sight view to the television.

Interestingly, this limitation is not so much a function of the mode such as AM and SSB, as it is the actual frequency you are using to send and receive signals.

Frequencies and Propagation

Previously in the Spectrum Regions and Services table above, you were asked to observe two characteristics.

- As wavelength increases, frequencies decrease.
- As wavelength decreases, frequencies increase.

Now let's update these characteristics with propagation effects.

- As wavelength increases, frequencies decrease and signals propagate through groundwave and skywave effect.
- As wavelength decreases, frequencies increase and signals can only propagate through line-of-sight effect.

The result is that with groundwaves, lower frequency signals will follow the contour of the Earth. Furthermore, groundwave and skywave effects occur simultaneously. With lower frequencies skywave signals bounce off the ionosphere and they are reflected back to earth. A term often referred to as "skip."

Conversely, with line-of-sight propagation, the frequency of the signal is high enough that it will penetrate the Earth's ionosphere, which contains a high concentration of ions and free electrons. In a real sense, line-of-sight signals pass through the ionosphere like a bullet.

Propagation Effect Tipping Point

Where is the tipping point between groundwave/skywave propagation and line-of-sight propagation? A reasonable conclusion would be the 6m band in the VHF spectrum region, which is why it is called the "magic band." Above 6m and starting in the 2m band of the VHF spectrum region is where line-of-sight only propagation occurs throughout the bands.

You may wonder, given that an unobstructed view is necessary with line-of-sight propagation, is any other form of long distance (DX) communication possible with the popular 2m and 70cm bands in the VHF and UHF spectrum region? Absolutely!

Remember, higher frequencies use line-of-sight propagation to pass through the ionosphere like a bullet. This is exactly why line-of-sight propagation is a powerful way to communicate across the face of the Earth with other HAMs using a network of bi-directional amateur radio communication satellites in low Earth orbit.

This will be discussed in detail later, but for now, let's see how the comparative differences between these three propagation effects will affect your signals intelligence efforts.

Propagation and Signals Intelligence

Signals intelligence is a term used by the government to describe the monitoring, interception, and interpretation of radio signals, radar signals, and telemetry.

Militaries and government agencies use encryption, coding and other techniques to prevent others from monitoring their communications traffic. However, by law, ordinary citizens are not allowed to transmit any coded or encrypted signals via any FCC designated frequency allocated for private use by citizens and HAMs.

What this means, is that every signal you send is sent "in the clear" which means that God and everybody can listen in. That by golly is the way it is.

Do not assume that using a privacy code with a local GMRS repeater or a PL tone with a local VHF or UHF repeater gives you privacy. All they do is to prevent you from hearing whoever else is on the same frequency when they are speaking. In other words, privacy codes make you and whoever else is using the same repeater frequency, selectively deaf – not eavesdroppers, unless you can manually break the squelch.

Once the enforcement capabilities of the FCC have been curtailed by the chaos of a global tribulation, survival communities will no doubt begin to employ several different encryption and coding methods to make their communications more secure. Meanwhile, there is no absolute way to make your communications secure.

However, there are three things you can do to help mitigate the ability of others to monitor your communications traffic: language, beam width and transmit power.

Using a Dead Language

The English language is the international language of Amateur radio. Not by law, but by convention. This means no HAM in any country is bound by any law to exclusively use the English language for identifying themselves and to transmit messages.

There is a long standing precedent for this. During WWI, Choctaw and other American Indian operators transmitted battle messages in their native tongue by landline telephones. Then in WWII, Comanches, Choctaws, Hopis, Cherokees, and other Native American Army radio operators transmitted battle messages over the air in their native tongues. Nicknamed the "code talkers" these men were highly courageous because the Japanese targeted them for capture and should that become a possibility, their own leaders would have had to kill them.

If you have several Native Americans in your community who are fluent in the same native tongue, congratulation, you already have a population of code talkers. If not, you can learn a dead language that is no longer in everyday spoken use but not just any dead language.

By definition, a language dies once it is no longer in everyday spoken use. The most widely known dead language is Latin. Everyone, who has been sold a raw deal and is then stuck with it, has learned the popular Latin phrase *caveat emptor*, "Let the buyer beware" the hard way.

From the ancient Latin language of the Romans has grown the modern romance languages which include: Spanish, Portuguese, French, Italian, Romanian, and Catalan. However, English is not a romance language. English is a Germanic language, specifically, from the West Germanic language group which also includes German, Netherlandic (Dutch) and Yiddish.

English, German and Netherlandic are living languages and difficult to learn. As a result of the Holocaust in WWII where 6 million European Jews were murdered by the Nazis, their principal language, Yiddish, became a dead language. Yiddish was the common every-day language and Hebrew was used for religious ceremonies.

Today, very few Jews in Israel, America and Europe speak this dead language. A 1996 report by the Council of Europe estimates a worldwide Yiddish-speaking population of about two million.

However, there is a historical tradition that is keeping Yiddish from becoming extinct altogether and this provides an opportunity for signals intelligence.

There are several good online sources for language learning recordings, books and tools for learning Yiddish today, which is lot easier to learn than Native American tongues. Yiddish is considered a Germanic language and closely related to High German. However, it is actually of Ashkenazi Jewish origin.

Yiddish is not unknown to Americans. We hear a few words of it each day such as "schmuck," a Yiddish word for a foolish or contemptible person. We also heard it with entertainers. Those old enough to remember popular tunes from the WWII era enjoy the pleasing sounds of the Andrew Sisters. In 1937 they began performing a Yiddish love song titled, *Bei*

Mir Bist Du Shein, which generally translates to "To Me, You Are Beautiful." It was a big hit then and is still popular with those who enjoy the music of that era.

More recently in 1974, producer Norman Lear formed the T.A.T. Communications Company. T.A.T. comes from the Yiddish "Tuchus Affen Tisch." While the polite definition is "Putting one's butt on the line, the actual Yiddish means "ass on the table." This is a Jewish version of the Christian admonition, "Jesus come quickly."

So, when other operators ask you to tell them the language you're using, they may also sense an untraceable familiarity.

However, no matter who asks, never tell them it is Yiddish. Tell them something like it is an ancient dialect brought to America by European settlers in the 1800s called Hoochie Coochie, or whatever else – besides Yiddish – that comes to mind. Then say no more.

As we live in an Orwellian age of correct speak, should learning Yiddish offend your sensibilities, at least try to apply the same evaluation framework to whatever language you wish. Whatever it may be, remember that America is a land founded by European immigrants.

Narrow Beam Width

Earlier we saw the difference between omnidirectional and directional antennas, with our cross-band repeater example. Instead of using a whip antenna with a beam width that equally propagates the signal in every directional a directional antenna such as a Yagi can narrow the beam considerably. The result is that eavesdroppers who are outside of the beam width of your directional antenna will have a very difficult if not impossible time trying to surveil your communications.

With lower frequencies in the MF and HF spectrum regions which propagate signals through groundwave and skywave, there is some beam width signals intelligence benefit.

However, higher frequencies in the VHF and UHF spectrum regions only propagate through line-of-sight. This is why they offer the best way to diminish the ability of terrestrial eavesdroppers outside of your beam width, to surveil your communications.

Transmit Power

A golden rule of amateur radio is to always use the least amount of power necessary to transmit a signal. With this in mind, imagine you are leading a field team beyond the outer perimeter of your community and you have a clear line of sight to a community cross-band repeater.

To help thwart eavesdropping on your signals, you're using a Yagi antenna with 90° beam width. To either side of your beam width, the signal will obviously be weak or non-existent. But what about directly behind you?

Assume that a predator is surveilling your communications and that they are behind your field team with a clear line of sight to their position and to the community cross-band repeater they are using. In this situation they will hear everything the community base station is transmitting to the field team via that repeater. That is assuming the signal has enough strength to travel far enough to reach them as a copyable transmission. In this case a directional beam width is not an effective solution. However, transmit power does offer you an advantage.

Regardless of how much power you transmit with, the cross-band repeater should only transmit with just enough power to reach you. Not beyond you. So, 15 watts of transmit power from the repeater barely reaches you, 50 watts will definitely paint a much large reception area behind you and the community's outer perimeter. Do not give this power to predators. Never!

With this we have covered all the technical concepts for you to be fluent when assessing your best possible options for any communication mission.

The Keys to the Kingdom

Yes, this chapter has re-hashed many concepts introduced previously in the book, but this was necessary for two reasons. Whether you are communicating by radio, writing a book, or whatever, as the communicator, the responsibility for being understood is yours and yours alone. If you have the temerity to tell someone that they are not bright enough to understand you, who really is the idiot?

The point here is that receivers are not responsible for understanding you. It is this way because the number one reason for communication failure is the assumption of it. So Dear Reader, please pardon me if I have laboriously re-hashed a lot of concepts you've already mastered.

However, after you read the table on the final page of this chapter, you will know all of your available options for 70% of your daily survival community communications needs. From this table, you can refine your search to narrow it down to the few options that best serve a particular communication mission requirement.

It is why this table, Survival Services and Phone Modes is going to be a key to the kingdom so to speak, and it adds something new, calling frequencies. HAMs use them to make initial contacts before moving the conversation to another frequency in the same band.

Survival Services and Phone Modes			
Channelized	**Amateur**	**Phone Modes**	**Calling Frequency**
Medium Frequency (MF) - 300-3000 kHz			
	160m	**AM, LSB**	
High Frequency (HF) – 3 to 30 MHz			
	80m	**AM, LSB**	3.885 MHz AM
	60m	**USB**	

Survival Services and Phone Modes			
Channelized	**Amateur**	**Phone Modes**	**Calling Frequency**
	40m	**AM, LSB**	7.290 MHz AM
	30m	**n/a**	
	20m	**AM, USB**	14.286 MHz AM
	17m	**AM, USB**	
	15m	**AM, USB**	
	12m	**AM, USB**	
CB / SSB CB (11m) (*SSB = LSB & USB)*		**AM, LSB, USB**	Channel 11 AM Channel 36 or 38 LSB
	10m	**FM, AM, USB**	29.600 MHz FM Simplex
Very High Frequency (VHF) – 30 to 300 MHz			
	6M	**FM, AM, USB**	50.125 MHz SSB
	2m	**FM, AM, USB**	146.52 MHz FM Simplex
	1.25m	**FM, AM, USB**	222.1 MHz SSB
Ultra High Frequency (UHF) – 300 to 3000 MHz			
	70cm	**FM, AM, USB**	432.10 MHz
FRS/GMRS (65cm)		**FM**	Channel 20 – Motorola Channel 6 – ICOM

If you are beginning to sense that everything you've learned since the first chapter was all a prologue to this table, you're right. This is because you can now read every entry in this table in a special way. Not to understand theoretical concepts, but rather, you can read it as easily as you would the aisle descriptions in a retail store. If you know a certain aisle in the store is where you need to look for what you want, that's where you go. After that, it is about narrowing down your choices. The same holds true here, because now you know the name of every aisle in the store.

Next up. Will a channelized expendable radio be one of your best options? To answer that Dear Reader, let's go shopping.

14

Channelized Two-Way Radios

Unlike amateur transceivers than can be tuned to a specific frequency with a variable frequency oscillator (VFO) a channelized transceiver uses channels that are preset for specific assigned frequencies. There are three channelized services you will use throughout the global tribulation and beyond, Family Radio Service (FRS), General Mobile Radio Service (GMRS) and Citizens Band (CB).

The FRS and GMRS share 22 channel presets with differences being in transmit power and GMRS repeater support. With CB, there are two variants. The basic CB, which only offers AM mode with 4 watts of transmit power, constitutes some 90% of all CB transceivers sold. The other variant, SSB CB transceivers, offer both AM mode with 4 watts of transmit power and SSB mode with 12 watts of power.

While channelized transceivers lack the robust features of more sophisticated amateur transceivers, the availability of channelized transceivers far exceeds the availability of tuneable amateur transceivers by orders of magnitude.

For this reason, regardless of whether or not you will use channelized transceivers with levels 1 and 2 of your community communication strategy, you will need them for level 3 command base station to maintain lines of communication with other local survival communities and municipalities. Therefore, for command base station use, at least two high-quality mobile transceivers for FRS/GMRS and SSB CB is highly advisable.

However, for low-risk level 1 and level 2 communications missions, channelized transceivers offer three significant advantages: ruggedness, simplicity and cost. The question then becomes when is it a good time to use a channelized transceiver instead of a more robust and expensive amateur transceiver?

When To Use A Channelized Transceiver

When it comes to the FRS/GMRS, CB, and SSB CB services, a title from a 1945 war film starring Robert Montgomery and John Wayne says it all – *They Were Expendable*.

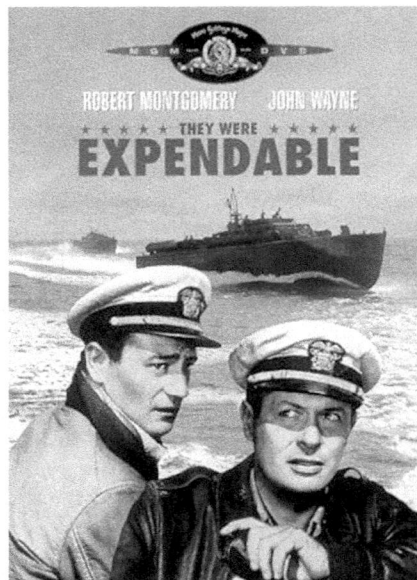

The dictionary defines expendable as: "of little significance when compared to an overall purpose, and therefore able to be abandoned." However, in terms of the coming global tribulation, a more suitable definition is: When something is deemed to be expendable, it can be sacrificed in order to accomplish a survival objective.

The operative word is "something" and not "someone". Your "someones" are not expendable. They are the future. What will be expendable with regards to your community communication strategy will be single band channelized two-way radios.

This is because they are widely available through numerous online and retail outlets and cost considerably less than more capable, amateur transceivers. Therefore, the criteria that drive the process of finding the best options are:

- **Mission Risk Profile:** Missions such as field operations beyond the outer perimeter of your community present a high-risk of death or injury. However, many missions such as those inside the outer perimeter of your community will present little or no risk of death or injury. Therefore, channelized transceivers, are well-suited to low-risk missions but should be choice of last resort for high-risk missions.

- **Phone Mode:** With Citizens Band (CB), you have a choice of 40 channels in AM phone mode plus another 80 channels in SSB phone mode (only with SSB CB transceivers). With FRS and GMRS, FM is the only available phone mode.

- **Mission Range:** Defining short range and long range with precise values is difficult, but a simple way to define range with channelized transceivers is with what you can or cannot see with a naked eye observation.

  Short: You can see a human figure without binoculars. Or the operator is nearby but there is a major obstruction to the line of sight.

  Medium: You can see the intended operator using binoculars or a rifle scope with a partial obstruction to the line of sight.

Maximum: You have a clear line of sight to the area the intended operator is in, but not the operator. Or, the operator is beyond your line of sight.

Use these ranges as initial approximate values. The refine them based on variables such as transceiver type, antenna, grounding, transmit power, terrain, weather and so forth.

● **Transceiver Type:** Mobile and Walkie Talkie transceiver types are available with both channelized services. From a practical standpoint, you'll want to stockpile FRS/GMRS walkie-talkies and SSB CB mobiles for low-risk missions inside and outside of your outer perimeter. For your command base station, a GMRS mobile transceiver is preferable.

After assessing the communication mission with a low-risk profile and which requires a phone mode, you have four channelized transceiver choices.

Service	Band	Mission Ranges and Phone Modes		
		Short	Medium	Maximum
CB (Basic)	HF (11m)	AM	n/a	n/a
SSB CB	HF (11m)	AM	AM / SSB	SSB
FRS	UHF (65cm)	FM	n/a	n/a
GMRS	UHF (65cm)	FM	FM	n/a

Whichever channelized transceiver you choose, never forget, these are the most numerous and easy to operate for those with little technical proficiency. Consequently, you're going to have many more people listening in to your communications traffic than you would with amateur service transceivers. For this reason, amateur transceivers are better for missions with a high security requirement, regardless of risk.

GMRS Transceivers

FRS/GMRS radios operate in the UHF band. Immensely popular with citizen enthusiasts, they offer ranges comparable with more robust UHF amateur transceivers while being extremely inexpensive in comparison.

The most numerous channelized transceiver type you'll likely use will be an FRS/GMRS walkie-talkie. Often referred to as "bubble pack" radios, as they are sold in multiples.

Even though they are simpler to operate than amateur transceivers, FRS/GMRS transceivers are ironically confus-

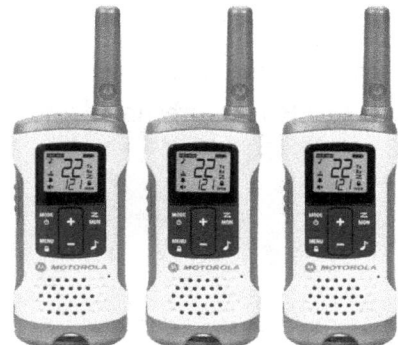

ing for many users for two reasons. An awkward channelized frequency allocation scheme by the FCC and the ability of manufacturers to use non-standard, proprietary configurations.

In terms of a survival, the main advantage of FRS/GMRS transceivers is the low cost of handheld FRS/GMRS walkie-talkie transceivers. Watt for watt, they are the least expensive handled transceivers available on the market today.

However, the cost of more powerful single band GMRS GMRS mobile class transceivers is no bargain. In fact, they cost nearly as much as a more feature-robust, dual band VHF/UHF mobile amateur transceivers. With this in mind, let's quickly review the key differences.

FRS/GMRS Frequencies

With amateur radios, hundreds to thousands of frequencies are available depending on what is allowed by the operator's FCC license. The extremely low-power FRS service does not require a license, but GMRS, does require a license (without testing) that will cover a whole family.

Consequently, when the FCC combined the unlicensed FRS and licensed GMRS services into a shared bandplan, they created a quagmire of twenty-two separate simplex channels. The most recent version of the bandplan was released on September 28, 2017 for these combined services and it is still a bureaucratic work in progress.

Another problem with the bandplan is that manufacturers can make this bandplan quagmire even worse, as they can assign different channels to different frequencies. For example, to make two different brands of FRS/GMRS transceivers work together, one may need to be set to channel 20 and the other to channel 21.

For this reason, the smart strategy for FRS/GMRS is not only to stockpile the same brand but the same model transceiver as well. The reason is, the way that a FRS/GMRS transceiver is used and the actual maximum transmit power it delivers for any given channel will vary.

With some models you can designate the service for a given channel and others may have factory defaults, so be sure to carefully read the user documentation for how channels, services and transmit powers are handled by that specific brand and model. In other words, look and look hard before you leap.

While there are many brands and models of FRS/GMRS walkie-talkies, only a small handful offer repeater support. Therefore, the principal use will be simplex mode without repeater support.

In the bandplan table below the three frequency ranges for shared simplex use with both FRS/GMRS are presented and the suitable mission ranges for each channel.

Shared Simplex FRS and GMRS Bandplan – as of September 28, 2017			
Channel	**Frequency**	**FRS Power (Watts)**	**GMRS Power (Watts)**
Medium Power (Medium Range)			
1	462.5625	2.0	5.0
2	462.5875	2.0	5.0
3	462.6125	2.0	5.0
4	462.6375	2.0	5.0
5	462.6625	2.0	5.0
6	462.6875	2.0	5.0
7	462.7125	2.0	5.0
Low Power (Short Range)			
8	467.5625	0.5	0.5
9	467.5875	0.5	0.5
10	467.6125	0.5	0.5
11	467.6375	0.5	0.5
12	467.6625	0.5	0.5
13	467.6875	0.5	0.5
14	467.7125	0.5	0.5
High Power Shared FRS and GMRS Simplex (Long Range)			
15	462.5500	2.0	50.0
16	462.5750	2.0	50.0
17	462.6000	2.0	50.0

Shared Simplex FRS and GMRS Bandplan – as of September 28, 2017			
Channel	**Frequency**	**FRS Power (Watts)**	**GMRS Power (Watts)**
18	462.6250	2.0	50.0
19	462.6500	2.0	50.0
20	462.6750	2.0	50.0
21	462.7000	2.0	50.0
22	462.7250	2.0	50.0

How you allocate FRS/GMRS channels will depend on the mission range and here are a few examples:

- **Low Power (Short Range):** Channels 8 to 14 assign the same 0.5 W transmit power to both FRS and GMRS. Since these services are in the upper UHF band at 65cm, the signal wavelength is short and can penetrate man-made structures better than VHF or HF frequencies. Consequently, these low power channels are ideal for short range, low-risk mission roles within the community.

- **Medium Power (Medium Range):** Channels 1 to 7 assign 2 W of power to the FRS service and 5 W to the GMRS service. These medium power channels are ideal for medium range, low-risk administrative and security missions within the community.

- **High Power (Long Range):** Channels 15 to 22 assign 2 W of power to the FRS service and 50 W to the GMRS service. These high power channels are ideal for maximum range, low-risk mission roles for field teams operating beyond the outer perimeter of your community. Remember, they are not advisable for high-risk missions.

If your community is on a tight budget and you cannot afford MIL-STD amateur transceivers for your field teams, then ruggedized FRS/GMRS walkie-talkies like the Motorola MS355R Talkabout discussed earlier is a good place to start.

GMRS Repeater Support

Repeater support is a good example of the FRS/GMRS bureaucratic quagmire. With FRS, repeater capability is not available on any channel. It only available with the GMRS service and in a limited role on channels 15 to 22.

The reason why GMRS transceivers offer a limited role is that only a few handheld transceivers such as the MS355R Talkabout GMRS walkie-talkie are repeater capable. Most GMRS mobile transceivers offer maximum GMRS transmit power and repeater support.

A good example of a GMRS mobile with repeater support is the Midland Radio, Micro Mobile MXT400. Unlike the MS355R, the MXT400 does not support FRS. It is a pure GMRS mobile which means it does not support the low power (short range) channels 8 to 14.

In terms of a viable level 3 community communications strategy, this is fine as the 40 W, Midland MXT400 is compliant with the FCC FRS/GMRS bandplan as of September 28, 2017 and you should equip your level 3 command base station with two of these GMRS mobiles.

In this case, it can be used on channels 1 to 7 for low risk, administrative and security missions within the community. Channels 15 to 22 are best for maximum range, low risk mission roles for field teams. However, a good advantage is that the MXT400 is well-suited for use with GMRS repeaters.

Like local repeaters used by amateur radio clubs, GMRS repeaters use two frequencies for each repeater channel as noted in the table below.

GMRS Repeater FCC Bandplan as of September 28, 2017			
Channel	**Output Frequency**	**Input Frequency**	**GMRS Power (Watts)**
RPT15	462.5500	467.5500	50.0
RPT16	462.5750	467.5750	50.0
RPT17	462.6000	467.6000	50.0
RPT18	462.6250	467.6250	50.0
RPT19	462.6500	467.6500	50.0
RPT20	462.6750	467.6750	50.0
RPT21	462.7000	467.7000	50.0
RPT22	462.7250	467.7250	50.0

The output frequency is the signal coming from a repeater. The input frequency is an offset for transmitting a signal to the repeater.

Like amateur repeaters that use a PL tone, the MXT400 GMRS transceiver offers 142 different privacy codes, which are similar in function and their use is optional.

Privacy codes are predetermined by whoever configures a GMRS repeater and then makes these talk path settings known. Here, we find a thorny issue.

Even though GMRS transceivers offer a large range of assignable privacy codes, the problem is that these codes are predefined by the manufacturer. For this reason, using privacy codes with dissimilar GMRS brands and models can result in interoperability issues.

This begs the question: are privacy codes a necessity for survival communities? The answer is a no. In the author's opinion, you should not use them at all. This is because privacy codes do not prevent others from hearing you or from transmitting on a given channel. Rather, they just prevent you from hearing the traffic of outsiders transmitting on the same channel.

GMRS Summary

As compared with amateur services, the biggest problem with GMRS is the lack of consumer off-the-shelf repeater solutions.

Radio associations across the country offer localized GMRS repeater services with commercial grade equipment. Therefore, getting one to work for a survival community will present a difficult and costly technical challenge and all that for just 8 repeater channels. Ergo, you will need to salvage an operational commercial grade GMRS repeater, or engineer a DIY (do it yourself) repeater with whatever you can scrounge or have on-hand.

The bottom line with GMRS repeater technology is that it works, but it requires expensive equipment that is difficult to obtain or build. Therefore, if you need to provide repeater support for field teams operating beyond your outer perimeter, amateur services are far superior to channelized services.

The bottom line with FRS/GMRS is that the FCC bandplan is a bureaucratic and manufacturing mishmash. If you really need to do FRS/GMRS, set your expectations low, because if you're going build a robust system, you will eventually find yourself in a bandplan quagmire as you try to make these inexpensive consumer transceivers perform tasks better suited to more expensive amateur and commercial equipment.

On the other hand, there is one affordable channelized service that has a simple, rock stable, any brand, and any model bandplan – Citizens Band (CB). That being said, FCC rules clearly state that repeaters and amplifiers are not legal for use on the CB band, so keep that in mind.

Citizens Band Communications

Regardless of the two way radio type, there are fundamental criteria such as cost, transmit power, range, availability and audience and each type has a different mix. In the coming global tribulation, CB two-way radios will offer a truly unique and valuable virtue – vast availability.

The CB radio service dates back to the 1960s, but it was in 1973 during the oil embargo that we saw an explosive growth in the use of CB radios. Unfortunately, with the advent of cell phones, interest began to wane. Consequently, all across the country, old 23-channel, 40-channel and SSB 40-channel CB radios can be found packed away in boxes stored in attics, garages and basements in every city and town in the country.

How many serviceable CB radios are packed away? That's hard to say but for grins, here is a conservative guesstimate. Consider this. 1976 was the peak year for CB sales with a record 11 million sold that year along. In 1977, another 9 million were sold and in 1978 another 8 million were sold. In other words, during a three year period, over 28 million CBs were sold. Wow!

If we ignore CB sales for all other years, and only focus on this three year period with an assumption that 10 percent of these are still packed away, that means there are possibly 2.8 million used two-way CB radios gathering dust, but how many of them still work?

Assuming that half are no longer serviceable, that's still approximately 1.5 million used, serviceable radios. While this figure is highly conservative, it is more likely that there are two to three times as many serviceable used CB radios. The question is where do you find them?

An interesting thing about Americans, is that they tend to hold onto old CB radios, even though they do not use them and nor understand how to assess their value. Ergo, if you like to haunt garage sales and know what you're looking for, there can be fabulous bargains to be had. Just be sure to test old radios dating back to the 1970s before you buy them, because if their electrolytic capacitors have dried out, they're only good for spare parts or as door stops.

Interestingly enough, finding used CB antennas is the problem, because the general assumption is that old radios are more valuable than old antennas. Consequently, many expensive CB antennas have been thrown away over the years. This is unfortunate because good, high-quality used antennas are often worth more than used CB radios.

However, CBs have a unique distinction among all types of two-way radios. Microphones, antennas and other CB accessories can be found in most every large truck stop, making them a wonderful resource.

Another interesting benefit with used CB radios is that folks will usually disconnect the antennas, power supplies, and microphones before they box them up. Disconnected from power sources and antennas, a good portion of these stored CBs will survive EMP and lightning spikes.

What this means for global tribulation survivors is that foraging for old and serviceable two way radios will yield productive results.

Yet, there is a question that is most often asked by those who do not personally own or use a CB radio. Are CB radios dead? Absolutely not.

Are CB Radios Dead?

In an age of smartphones, driver apps and GPS navigation are CB's still useful? According to conservative estimates, over half of all long haul truckers in America still use CB radios. Sure, truckers no longer need to rely on CB radios to do everything, but then again, they keep using them because nothing else does exactly what a CB does, such as:

- Group voice conversations

- No cellular coverage gaps
- Local weather and road conditions
- Speed traps, traffic and detours
- No monthly calling plan fees

The range of an AM CB radio is under five miles unless atmospheric conditions promote skip. (Skip is a term to describe ionosphere skywave propagation.) However, the limited range of a CB has more to do with the AM phone mode and FCC restrictions on transmit power.

Nonetheless, during the global tribulation the absence of FCC regulation will result in a large number of CB owners finding ways to amplify their transmit power levels so as to extend their range and voice quality. With this in mind, the relevant global tribulation CB pros and cons are:

Comparison Category	CON	PRO
Channelized Service	A limited number of channels preassigned for this service by the FCC and manufacturers.	Each channel is paired with a specific frequency. This standardization allows for consistent use internationally.
Operating Modes	Over ninety percent of all Citizens Band radios are single mode radios. That being amplitude modulation (AM).	10 percent of all Citizens Band radios offer two operating modes. AM and Single Sideband (SSB)
AM Modulation	AM is the least efficient mode for transmitting a signal. Consequently, the AM range of CB radios is limited.	The groundwave propagation characteristics of AM make it ideally suited for mobile to base station communications.
SSB Modulation	Using a SSB clarifier to tweak a CB signal from other operator can be distracting with mobile communications.	As opposed to AM, SSB is the most efficient way to transmit signals locally, nationally and internationally.
Number of Channels	Prior to 1977, CB radios were limited to 23 AM mode channels. Current models offer 40 channels.	With dual mode CB radios 40 channels are available with AM mode and another 80 channels for SSB mode. This makes a total of 120 channels.
Interference	At present, SSB mode operators limit themselves to channels 16 and 36-40, to avoid AM mode interference.	During the global tribulation, SSB operators will use whatever is available as primary channel users.
Transmit Power	AM mode range is a few miles due to a limit on transmit power to just 4 watts. The use of amplifiers to increase to transmit power is expressly	A CB in SSB mode offers a transmit power of 12 watts. This added power plus the high efficiency of this mode makes it ideal for long-

Comparison Category	CON	PRO
	prohibited by law.	distance base station operations.

As noted in the table above, less than 10% of all CBs are SSB CB. The rest are standard basic CBs with AM phone mode only. For command base stations operations, SSB CB mobile radios are the ticket. They offer excellent range with low-power and low cost.

SSB CB for Base Station Operations

Previously, the Bearcat 980SSB was presented as a good example of SSB CB radio with good ergonomics and a simple user interface for field operations.

Bearcat 980SSB CB 11 M			
Input Power (VDC)	Current Drain Amps (A) Milliamps (mA)		Transmit (Watts)
VDC Input Range	Receive Drain	Max Transmit Drain	Maximum Output
13.8 V DC	650 mA	4 A	12 PEP

For command base station operations, at least two CB radios are advisable for 24/7 use. One dedicated to AM mode and the other to SSB mode. You'll also want to ensure the best possible performance of these transceivers with professional tuning and aftermarket components.

With this in mind, here are four things you can do to optimize your SSB CB transceivers:

- **Radio Tuning:** CB radios (especially older used models) need to be fine-tuned by a competent technician using proper test equipment. This requires removing the CB cover to access the tuning devices built in to the radio, to ensure the best performance of the transceiver. Remember, smart operators have their CBs professionally tuned.

- **Solid State Mobile Amplifier:** A DC-powered amplifier designs for mobile and base station applications is preferable. A black market amplifier with 50 W to 100 W of transmit output power will cost roughly the same as a new SSB CB.

  Keep in mind amplifiers do not boost the signal strength of incoming traffic from another operator unless equipped with a preamplifier. Therefore, while these other stations will hear you more clearly, the same will not hold true for you as you may be unable to hear them clearly or at all.

- **Receiver Preamp:** Several mobile solid state mobile amplifier models come equipped with a preamp to improve the sound quality of incoming traffic. Keep in mind, there are concerns with preamps. For example, a preamp can help to filter out the distortion and noise of a weak signal, but they can also boost the noise level as well. Therefore, as a general rule of thumb, the closer a preamp is to the antenna the better it will work. Make sure your CB system has been properly tuned before using it with a preamp.

- **Base Station Microphone:** The use of a good voice quality base station microphone such as a Shure 522 will give you a stronger and much clearer on air sound than factory-supplied PTT hand microphones.

This brings us to a question often asked by SSB CB operators, especially those with an FCC amateur license. If you can get excellent performance from an SSB CB system for just a few hundred dollars, why spend thousands on an amateur HF system?

This question is legitimate because as was pointed out earlier, CB radios only work the 11 meter HF band, which means you have the benefits of HF but within a narrowly defined limit. We'll address this in the next chapter.

Chapter Summary

Channelized transceivers are widely available and generally much less expensive than amateur and commercial components. However, this comes at a price. Flexibility. With GMRS you can use a repeater, assuming you can build one and assuming your mix of brands and models will work together.

On the other hand, repeaters are not available for the CB and SSB CB services by law and the 4.0 watt maximum transmit power for AM mode gives these two-radios a short leash when it comes to range.

If you have properly assessed your needs and can use a channelized service, you will save money, assuming you have set proper expectations. If you want to do more and expect to, then you will need to continue your search for the best options for use in amateur two-way radios.

15

Amateur Two-Way Radios

In the previous chapter, we explored the use of channelized transceivers, FRS/GMRS, CB and SSB CB for missions requiring a phone mode, AM, FM or SSB. However, amateur transceivers also support these same phone modes, plus more.

As we stated previously, during the global tribulation, 70% or more of your radio traffic will be via a phone (voice) mode, across a broad range of spectrum regions, services and modes. The remaining 30% will be other modes such as CW (Continuous Wave – Morse code), RTTY, Data and so forth.

As there is an obvious overlap between channelized and amateur transceivers, this chapter will principally address the other 30% of your radio traffic requirements. These requirements will be more sophisticated than the phone modes we've discussed to this point. They will require additional specialized components and experienced operators.

However, these advanced capabilities are only available with amateur transceivers which can do many things that are unavailable with channelized transceivers, but which will still prove to be decisive. These non-phone modes are: CW, RTTY, Data and Image.

These modes are available with most modern VHF/UHF and HF amateur analog transceivers and with an interesting twist.

There is one thing you can do with VHF/UHF amateur transceivers that is impossible for channelized transceivers. Extremely long distance communications (DX) range in the FM mode. An advantage that may prove useful for field teams operating beyond the outer perimeter of your community. But first, let's see how you can use a modern analog VHF/UHF amateur transceiver in your command base station in a very effective way.

VHF/UHF Communications

The VHF and UHF spectrum regions will be the primary workhorses for local low-risk and high-risk missions beyond your outer perimeter. As was discussed earlier, the excellent open terrain characteristic of VHF spectrum region makes it ideal for supporting field operatives.

For communication missions in areas with man-made structures, the superior penetration characteristic of UHF signals is where both FRS/GMRS and UHF amateur transceivers have something in common. UHF signals are well-suited to missions where your signals must penetrate reinforced concrete structures.

However, for level 3 command base stations in the VHF and UHF amateur spectrum regions, the Kenwood TM-V71A is a real workhorse. A very rugged dual band (VHF/UHF) 50 W amateur transceiver, it is ideal for high-risk operations on both sides of your outer perimeter.

Kenwood TM-V71A

For your level 3 command base station, a dual band transceiver like the Kenwood TM-V71A is going to give you some very flexible options, thanks to its true dual band mode.

Kenwood TM-V71A – 144/430 Mobile Transceiver		
Input Power (VDC)	Current Drain Amps (A) Milliamps (mA)	Transmit (Watts)
VDC Input Range	Receive Drain	Max Transmit Drain	Maximum Output
13.8 V DC ±15 %	1.2 A	13 A	50 Watts

When operating in dual band mode, the transceiver offers two band displays, one for Band A and the other for Band B, for the two available operating modes: transmit and operating.

- **Transmit Mode:** Also called the "control" mode. Monitor incoming traffic and transmit on the same frequency.
- **Operating Mode:** Monitor incoming traffic only.

When assigning Band A to VHF and Band B to UHF, you can simultaneously monitor both bands. Likewise you could configure both Band A and Band B with different frequencies on the same band. In either case, you will hear traffic on Band A and Band B simultaneously.

This true dual band feature is so useful, that you will find two 3.5mm audio jacks on the back of the transceiver. One assigned to Band A and one to Band B. This allows you to set up your command base station TM-V71A with two external speakers. The best configuration for this dual external speaker setup is to position one speaker on the wall to your left as you face the transceiver and the other on the wall to your right. This way, as you monitor both bands, the directionality of the sound will aurally clue you into which band you're hearing the traffic on.

This raises a necessary question. Do you use one TM-V71A transceiver to simultaneously monitor two bands, or do you follow a dedicated band strategy using two TM-V71A transceivers? A big advantage of using two transceivers is redundancy. Also, you can alternate the transceiver configurations to verify that both are working properly.

However, when using a single, true, dual-band transceiver such as the TM-V71A, there is an easy way to use calling frequencies.

Previously, we identified the national FM simplex calling frequencies for the amateur VHF and UHF bands. VHF is 146.520 MHz and UHF is 446.000 MHz. With this in mind, let's set up our two TM-V71A transceivers to work these bands for a command base station.

- **TM-V71A No. 1:** The default for Band A and Band B is both are configured for VHF with the 144 option. Band A is assigned the calling frequency, 146.520 MHz. It will be your default transmit and operating band. Band B is set to operate your Tactical VHF frequencies.

- **TM-V71A No. 2:** The default for Band A and Band B are both configured for UHF with the 440 option. Band A is assigned the calling frequency, 440.000 MHz, and designated as your default transmit and operating band. Band B is set to operate your tactical UHF frequencies.

In both cases, each transceiver is configured for the memory mode – not the VFO (tuneable) mode. This way, if you have ten tactical channels selected for your field operations in each band, these frequencies will be stored in your transceiver's memory and easy to access.

Band A will be your default for operating and transmitting on the assigned national calling frequency. When you hear traffic that you need to respond to, make the contact and instruct the other operator to move to a different frequency on the same band. This is when you select Band B as the default for operating and transmitting band and dial in the alternate frequency to continue the contact via tuneable VFO or memory preset frequency.

Granted, using two TM-V71A transceivers for operations is more expensive, but this configuration offers more than enough options to compensate for the added costs to include:

- **Dual Band Transmit:** You can monitor and transmit on both the VHF and UHF bands simultaneously for two different missions, as opposed to juggling back and forth between bands with a single transceiver in order to serve two missions.

- **Operational Redundancy:** With two dual-band transceivers, if one unit fails, the other can be used to continue transmissions while the failed unit is repaired or replaced. To make this work best, the same VHF and UHF frequencies need to be stored in memory, so they are identical on both transceivers.

- **Memory Channels:** GMRS offers 8 channels with 50 W of transmit power. With SSB CB five channels (36 to 40) are typically used for SSB at 12 W. To really see how limited channelized transceivers are, consider this: The TM-V71Ahas 200 built-in memory settings you can program for any combination of VHF and UHF frequencies!

- **Frequency Programming Software:** Because amateur transceivers can work both bands, a wide number of frequencies, frequency programming software such as the freely-available public domain Chirp program can be used with a special cable connection between your computer and your transceiver, to populate the 200 memory channels with far less effort than is required manually.

Chirp works with the Kenwood TM-V71A mobile, the Yaesu FT-60R HT and many other brands and models. You use it to create a master spreadsheet with the frequencies you want to store. You can then upload the master for each of the M-V71A transceivers in your system.

Furthermore, the Yaesu FT-60R HT and other popular bands also offer a cloning cable. Once you're programmed your first FT-60R you can connect it to another FT-60R and clone all of the settings to the new transceiver. So when evaluating amateur transceivers, be sure to see if you can program it with a radio frequency program or a cloning cable. This will be a huge help in the long run.

Now, for the sake of discussion, let's assume you've found two old, single band 25 W mobile transceivers, one for VHF and one for UHF. Together, they cost you $50 which is a great bargain but can they be made to replace a single TM-V71A costing seven times as much, to do something useful that no channelized transceiver ever could? You betcha and it will leave you humming the song, *Ground Control to Major Tom* by David Bowie.

Amateur Satellite Service

You may wonder, given that an unobstructed view is necessary with line-of-sight propagation, is any other form of long distance (DX) communication possible with the popular 2m and 70cm bands in the VHF and UHF spectrum region? Absolutely!

Remember, higher frequencies use line-of-sight propagation to pass through the ionosphere like a bullet. This is exactly why line-of-sight propagation is a powerful way to communicate across the face of the Earth with other HAMs using a network of bi-directional amateur radio communication satellites in low Earth orbit.

Amateur radio satellites are built for and used by amateur radio operators to enable long distance (DX) communications between amateur radio stations. While few outside of amateur radio know about these satellites, their use dates back to the early 1960s.

Most of these are called OSCAR which is an acronym for Orbiting Satellite Carrying Amateur Radio. These satellites operate in a number of modes including FM phone, CW (Morse code), and data. Presently, these satellites are being built and/or funded by members of AMSAT, the Radio Amateur Satellite Corporation and AMSAT affiliates.

To get started all you need is a technician class license and linking up with these satellites is a reliable process as they have timely orbits. When a satellite is within range, the bands are open. It is that simple, so when you hear it, you know you can begin working it in the clear.

During the global tribulation, many of these satellites will remain operational because satellites are designed for the rigors of space, including solar storms. It is a safe assumption that some or many may will fail, but a good number should continue working for years, if not throughout the entire global tribulation and beyond. This will be important because at approximately 3,000 miles in diameter (depending on the satellite's height), their transmit footprint can be considerable.

An example of an OSCAR satellite in low Earth orbit that offers simplex FM phone mode service is UO-14:

Satellite Designation	Downlink Frequency	Uplink Frequency
UO-14	435.070 MHz	145.975 MHz

As you can see in the table above, UO-14 like other OSCAR satellites use the VHF band for the uplink (ground to spacecraft) frequency and UHF downlink (spacecraft to ground) frequency.

When the satellite passes overhead, the pass predictions for this satellite will vary. For example, the pass predictions for Reno, NV range from roughly four to fourteen minutes. To learn more, visit the www.amsat.org website.

With this in mind, let's configure a Kenwood TM-V71A mobile for FM phone mode DX communications with satellite UO-14, (declared dead in 2003.)

- **Band A:** Set to 145.975 MHz, Band A is used as the transmit (uplink). When you press the PTT button your microphone, your signal will be sent via this VHF frequency.

- **Band B:** Set to 435.070 MHz, Band B is used as an operating frequency. You can monitor the satellite's downlink transmissions via UHF.

In terms of power, HAMs can use transmit powers as low as 5 watts but for using FM phone mode, a transceiver with a transmit power of 25 watts or more is advisable.

As it so happens, the two used FM simplex transceivers you found as at a local garage sale as described above are old and do not offer a the PL tone capability required by most local amateur club repeaters. These are oldies but not necessarily goldies. However, what you will have are two simplex single band 25 watt transceivers, one for VHF and one for UHF. Now all you need are two medium beam directional antennas, one for each band, which is why circularly polarized crossed-Yagi antennas are popular with HAM satellite operators.

Granted, you can work satellite traffic with a single Kenwood TM-V71A mobile and run it on approximately half as amps as the two used transceivers, but it will cost you a lot more. On the other hand, satellite communications is a great way to breathe new life into a transceiver with limited functionality or which has failed partially, but is still useful on a more limited basis.

If you are wondering about the value of using satellite communications for a field team operating beyond your outer perimeter, satellite communications is a useful DX option. It is one that can augment FM repeaters and Single Side Band (SSB) for hopping ridgelines.

This brings us to a question often asked by SSB CB operators, particularly those with an FCC amateur license. If you can get excellent performance from a SSB CB system for just a few hundred dollars, why spend thousands on an amateur HF base station?

This is a legitimate question as CB radios only work the 11 meter HF band. This means that CB gives you the benefits of HF but only with a handful of channelized frequencies as opposed to the hundreds of frequencies available with amateur HF transceivers.

So why spend more for a good amateur HF transceiver? One answer is pizza. Yes, pizza. Visit a local pizzeria and what will you see? They sell their pizza by the slice as well by as the whole pie. Assuming the pizza is called HF, what you're buying with a SSB CB transceiver is a slice of the pie, which mind you, is every bit as tasty as the whole pie.

However, if you buy the whole HF amateur transceiver pie, not only will you get more slices, each slice will have its own unique toppings (in this case, modes.)   Let's see what you get when you buy the whole HF pie for your level 3 command base station.

HF Operations

Up to now, this book has principally discussed CB, with only two modes, AM and SSB. For HF amateur radios, the available modes are AM, FM, SSB, CW, RTTY, DATA, and Image. Therefore the most important differences between CB and amateur HF are as follows:

- **Frequencies:** With the amateur HF bandplan you're not confined just one HF band, as is the case with the CB the 11-meter band. Nonetheless, you can listen to any amateur band including HF bands without a license. You only need a license to transmit.

- **Licensing:** A license is not required to operate a CB or SSB CB radio, which is why the 11-meter band is not supported by amateur HF transceiver manufacturers. However, the FCC does require a license for amateur HF bands and frequency access will depend upon the class of license you hold and the country in which you operating.

- **Transmit Power:** The maximum transmit power for SSB CB is 12 W, whereas licensed amateur HF allows for a maximum of up to 1,500 W on most on most bands or sub-bands for operators with a general class license or higher and the maximum transmit power for a technician class amateur is 200 W or less.

- **Filtering:** Cleaning up the signals from incoming traffic is limited to squelch, noise blankers, and clarifier controls with SSB CB radios. To further improve the sound, CB operators will sometimes use preamps to increase the signal to noise ratio of an incoming signal. However, with amateur HF transceivers, the range of available signal processing and filtering features is considerable and will depend on the type and price range of your HF transceiver.

When evaluating HF transceivers for your command base stations, you will see four basic types designed for the HF and the 6m (50 MHz) band of the VHF spectrum region. Let's quickly compare the four.

HF Base Station

With a fully featured base station, you have transceivers with 100 W to 200 W and all of the possible filters and features available. These transceivers are used by HAMs who like to compete in international DX contests and can receive and clarify even the weakest long distance signals. Designed for fixed base operators, they are typically powered with 110/220 V AC power. This AC power requirement makes them unsuitable for operations in the field.

An example for a HF base station is the 200 W Kenwood TS-990S HF/50MHz transceiver. With a retail price of $7,979.00 (which does not include must have accessories), it provides the ultimate contest / DX experience. However, for the purpose of survival communication missions, it's a big white whale.

HF Portable

A portable HF transceiver is smaller than a base station but larger and more robust in terms of features than a mobile HF. They are designed for mobile operators operating from vehicles and are powered by 13.8 V DC. For command base station use, they can be powered from an external AC power supply or batteries.

Popular variants will have a typical transmit power of 100 W and will come with built-in automatic antenna tuner. The most popular HF portable on the market today is the 100 W Kenwood TS-480SAT, it is ideal for level 3, command base station communication missions.

Although it has fewer filters and features than the TS-990S (which costs 7X more) it has what you actually need to competently work the HF/50 MHz bands with 13.8 V DC and is well-suited to working non-phone modes such as CW, RTTY, Data, and Image.

All things considered, the Kenwood TS-480SAT represents the ideal type transceiver for survival command base station communication missions.

HF Mobile

Smaller than a portable HF, a mobile HF is ideal for operators operating from vehicles or on foot with a backpack and they are powered by 13.8 V DC.

Lacking many of the features of a portable or base station variant, a mobile HF will give you the necessary basics for working the HF spectrum region and HF/50 MHz bands. On the other hand, an HF mobile will have a smaller form factor (dimensions / weight) than a portable HF transceiver such as the Kenwood TS-480SAT, which is not recommended for backpack deployments beyond your outer perimeter.

A good example of an HF mobile that is suitable for backpack deployments beyond your outer perimeter is the Yaesu FT-891. It

offers a small form factor and with a retail price of $679.00, this HF/50 mobile has a lot to offer. Unlike the Kenwood TS-480SAT, it does not offer a built-in automatic antenna tuner but it does have two useful antenna options.

First is the Yaesu ATAS-25 manual tuning antenna system. It was designed by Yaesu for the FT-891 and the FT-817ND backpack field radio discussed previously and supports the: 40m, 20m, 15m, 10m, 6m, 2m, 70cm amateur bands. The other option is the Optional FC-50 Antenna Tuner Unit with a retail of $329.00.

This optional FC-50 Antenna Tuner Unit literally doubles the form factor of the FT-891, which makes it this combo suitable for survival command base station use only. Then there is cost. The combined cost for a Yaesu FT-891 with an external FC-50 antenna tuner unit is $1008. Not good when you compare that with the retail price of $1099.95 for the Kenwood TS-480SAT.

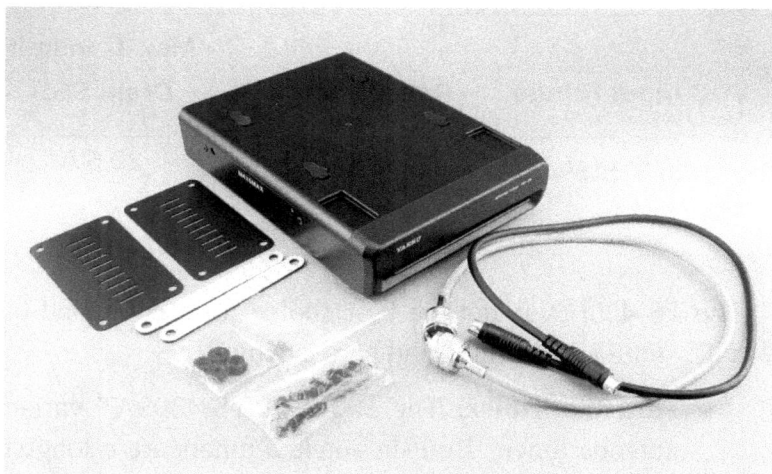

HF Handheld

Discussed extensively in the chapters dealing with low-power, level 1 QRP CW and level 2 SOTA class HF transceivers. Powered by 13.8 V DC they offer transmit power levels comparable with that of an SSB CB at 5 to12 W.

In terms of survival command base station use, HF handhelds run afoul of the Goldilocks effect. They'll either be too cold or too hot, but never, just right. However, keeping a level 1 QRP CW transceiver kit on hand with the base station is always advisable.

Therefore, in terms of price, features, and suitability for command base station communication missions, the Kenwood TS-480SAT is clearly a reliable benchmark HF transceiver.

Kenwood TS-480SAT

Kenwood makes two variants of the TS-480, the 100 watt TS-480SAT and the 200 watt TS-480HX.

Kenwood TS-480SAT HF/50MHz			
Input Power (VDC)	**Current Drain Amps (A)** **Milliamps (mA)**		**Transmit (Watts)**
VDC Input Range	**Receive Drain**	**Max Transmit Drain SSB**	**Maximum Mode Output**
13.8 V ±15%	1.5 A	20.5 A	100 W – SSB/CW/FSK/FM 25 W – AM

The TS-480HX is not the best choice for a command base station transceiver, but the 100 watt TS-480SAT variant is and here is why:

- **Antenna Tuner:** The 100 watt TS-480SAT variant comes with a built-in automatic antenna tuner. Built-in antenna tuners are a long-standing practice with military field and mobile transceivers and they work well with antennas pre-tuned for the transceiver. The 200 watt TS-480HX requires an external manual antenna tuner, which allows the use of untuned antennas, but this means more components, complexity and cost.

- **Cost:** The average retail sales price for the 100 watt TS-480SAT with a built-in antenna tuner is roughly twenty percent less than a 200 watt TS-480HX. But considering that a high quality external antenna analyzer and tuner can costs as much as the TS-480HX itself. Therefore, the TS-480HX becomes a very pricey option and this makes one wonder, is this extra cost worth the transmit power?

- **Power Source:** The 100 W TS-480SAT only requires one, 13.8 V power source. However, to transmit at 200 W with the TS-480HX, you'll need two, 13.8 V power sources.

The biggest difference between the TS-480SAT and the TS-480HX is how much amperage you need to feed it. The TS-480SAT only needs a single 25A power source, whereas the TS-480HX requires two 25A power sources or a single 50A power source.

Ergo, is going from 100 W to 200 W worth twice the power drain? A common misconception is that by doubling the transmit power from 100 W to 200 W, you will double the range. Wrong. The Inverse Square Law insures that.

Several factors determine range – not just transmit power alone. These factors include frequency, time of year, time of day, atmospheric conditions, antenna design, feedlines, and yes, transmit power. Consequently, when keeping the need for redundancy and the costs in mind, the higher cost of the 200 W capability of the TS-480HX is actually a negative factor.

Aside from antenna tuners, two other important HF considerations are frequencies and modes.

ITU Region 2 HF Frequencies

Even though the frequency designation used by Kenwood for the TS-480SAT is HF/50MHz, technically speaking, it is a three spectrum region MF/HF/VHF transceiver with eleven bands.

The product description covers the 6m (50MHz) band at the bottom of the VHF spectrum plus all the bands in the HF spectrum. However, it fails to mention the 160m band in the medium frequency (MF) spectrum. Given that few HAMs experiment with the 160m band, this is understandable.

As for the actual frequencies that you can transmit on, HF transceivers sold in America must work with an assigned bandplan. These bandplans are defined by the International Telecommunication Union (ITU) regions 1, 2 and 3, and each country within that ITU region, can further define the frequencies available with their local bandplan. In America, this is done by the FCC.

The TS-480SAT transceivers sold in the American market are configured for ITU Region 2 and cover all amateur bands from 1.8Mhz to 54Mhz as shown in the table below.

Spectrum	Band	ITU Region 2 Frequencies
Medium Frequency (MF)	160 m	1.800 MHz – 2.000 MHz
High Frequency (HF)	80 m	3.500 MHz – 4.000 MHz
	60 m	5.3515 MHz – 5.3665 MHz
	40 m	7.000 MHz – 7.300 MHz
	30 m	10.100 MHz – 10.150 MHz
	20 m	14.000 MHz – 14.350 MHz
	17 m	18.068 MHz – 18.168 MHz
	15 m	21.000 MHz – 21.450 MHz
	12 m	24.890 MHz – 24.990 MHz
	10 m	28.000 MHz – 29.700 MHz *
Very High Frequency (VHF)	6 m	50.000 MHz – 54.000 MHz *
*Can Be Used in FM Mode		

During the global tribulation, natural disasters, space weather and an unpredictable environment will combine to wreak havoc on one or more different segments of the entire radio frequency spectrum, so having a multi-band, multi-mode HF/50 MHZ transceiver is a must.

For this reason, the ability to work 11 different bands with an HF mobile transceiver like the TS-480SAT will give you the widest possible range of options, as each band has its own unique characteristics. All the rest the TS-480SAT can do on top of that is gravy and there is a lot of it.

Before moving into the different modes and the 9 bands in the HF spectrum, let's take a quick moment to look at two frequency oddballs so to speak, 160m and 6m.

160m – The Top Band

Known as the "Top Band" or "Gentleman's Band", the 160m band is located in the medium frequency (MF) spectrum region. The lowest radio frequency band allocated for use in most countries. It also has the distinction of being the oldest amateur band in use today. A general class license or higher is required to work the Top Band.

Today, the 160m band is principally worked at night by highly dedicated experimenters. During the summer months, you can bring in 5 W signals from hundreds of miles away, but noise can be a problem. However, during the winter months, you can bring in a 5 W signal from thousands of miles away with a relatively clean sound.

In terms of the global tribulation, the downsides with the Top Band are that it is dominated by a small population of experimental HAMs and it requires a very long antenna. To put this in perspective, a half wave 160 m dipole antenna needs to be 80 meters long (260 feet) which is why many HAMs operate this band with a shorter quarter wave antenna with a length of 40 meters (130 feet) on this band.

Regardless of whether you're working the Top Band with a half-wave or a quarter-wave antenna, that's still a lot of wire. Then again, if you're a farmer or ranger with lots of old electrified fence wire lying around and a top of the line antenna tuner, you're good to go.

Therefore, while you can work 160m with a TS-480SAT, the real issue is not so much about the antenna, but how many other operators you can make contact with on this band, especially when other HF bands are better suited to survival communications.

That is, except for one oddball band in the VHF spectrum region. This is one that military, government, and amateur operators truly appreciate. So much so, they call it the "magic band."

6m (50MHz) – The Magic Band

As stated previously, the frequency designation used by Kenwood for the TS-480SAT is HF/50MHz, which does not include the 160 m band in the medium frequency (MF) spectrum. However, the designation does include the 6m band (50MHz) which is in the VHF spectrum.

The reason HAMs call the 6m band the "Magic Band" is that it is like a box of chocolates. You never know what you'll get next, other than a lot of fun, which is why HAMs enjoy working the band. The principal reasons why are:

- Any HAM with a technician class license or above can use a transmit power up 1,500 watts.

- It gives operators the FM voice quality of the 2m band.

- The signal propagation with the VHF 6m band is similar to the bands in the upper HF spectrum.

- A VHF 6m half wave wire antenna will be a shorter than those required for HF bands.

It is interesting that Vietnam era PRC 25 & PRC 77 field transceivers used by our military featured two bands, including the 6m band. These older field radios were so efficient that even though they transmitted voice signals with just 1-2 watts of power, they had a range of 3 to 5 miles and could operate for 60 hours on a single battery. Legendary for their ruggedness and performance, small militaries around the world still use these venerable transceivers to this day.

No wonder, General Creighton Abrams, once called the PRC 25 "The most important tactical item in Vietnam today." The US military upgraded it to the PRC-77 variant which introduced voice encryption capabilities.

Nonetheless, the North Vietnamese were so impressed with the PRC 25 they adopted it for their own use and happily snatched abandoned transceivers on the battlefield. They became so important to the North Vietnamese, that China began making battery packs for them.

While our military moved away from 6m field transceivers decades ago, the 6m band is still in use today by municipalities who cannot afford digital transceivers or are in the midst of a transition to them. What is important to remember, is that the FCC has assigned special frequencies on the 6m band for government users and amateurs cannot transmit on them.

For municipalities that have switched to digital, the extreme environment of the global tribulation, may force some to return their old 6m equipment to service as infrastructure fails.

In cases where you develop a strategic alliance with a municipality using a 6m system, they can provide you with a transceiver or arrange a commonly available service.

With regards to the other HF bands supported by HF mobile transceivers such as the Kenwood TS-480SAT, each band has its own unique local and DX traffic characteristics and issues. These performance characteristics depend upon a number of factors to include:

- **Time of Day:** Some bands are better for day traffic, others for night traffic and a few that perform equally well.

- **Hemisphere Season and Time of Year:** The seasons are reversed between the Northern and Southern hemispheres of our planet. If you are in America and looking to make contacts south of the equator, the season of the year for the region you want to contact will determine the most suitable band.

- **Sunspot Activity:** The sun has an 11-year cycle with a period of minimum sunspot activity and maximum sunspot activity. During the last solar cycle, sunspot activity was inconsistent and going forward, we can expect more of this, as well as big solar storms.

- **Congestion and Service Noise:** Some bands are naturally more popular than others. Case in point is the HF 20m band. A day and night favorite of HAMs it offers good propagation and the ability to make lots of local and DX contacts, but is more prone to traffic congestion due to multiple operators working the same frequencies. As the FCC assigns multiple services to or close to these bands, this also generates noise on the band.

While there is a bounty of amateur HAM knowledge on HF band performance on the Internet, in the final analysis, finding what works best during a global tribulation will become a process of trial-and-error to identify which frequencies perform best in an adverse environment.

You may wonder. Why bother with making contacts with someone in Australia or Brazil? What could they possibly offer that would be of use? Plenty!

By reaching out across the globe with your HF transceivers, your HAMs will be able to gather valuable intelligence about what is happening elsewhere in the world and how survivors are dealing with survival issues.

Not only can you learn how the elites are shifting tactics elsewhere in the world, but you can also share valuable human knowledge such as how to make a natural antibiotic for treating ash boils for example. Imagine this. Save just one child's life with this kind of information and nobody will ever ask you again, "why bother with amateur HF?"

The other strong suite with a portable HF/50MHz transceiver such as the TS-480SAT is modes.

Mode Dependencies

Up to this point, we've focused on AM, FM and SSB phone modes. Now let's look at the other modes of HF operations, namely CW, RTTY, Data and Image. While these other modes are used for a wide variety of tasks and in different ways, what they all have in common is that you need external components to make them work. In other words, these modes require device-dependent extensions of your transceiver.

Tried-and-true analog HF/50MHz transceivers like the TS-480SAT do not have built-in computers, but have a robust range of connectivity with external devices such as computers, terminal node controllers (TNC) and external CW keyers.

This is important, because a true solution is something that works the way you need it to work, straight out of the box. Anything else can morph into a frustrating, expensive and unproductive fishing expedition.

As the modes supported by TS-480SAT have already been explained, this chapter will focus on the external devices needed, to make use of its connectivity features. During the course of the global tribulation, your community's scroungers may come across some old HF transceivers. When they do, they most likely will not be able to power them up and test them, nor will they have the time to do so. Rather, they'll need to eyeball the transceiver so as to make a quick on-the-spot decision. Here is what they'll need to look for:

- **Computer:** All data modes such as PSK31 will require a computer. Newer model laptops are the best choice as they have good quality sound cards, the latest USB version and can be powered directly from a DC source. For Windows-based systems, the author's recommended version is Windows 7 Professional 64 bit. For Linux-based systems, Debian is the author's recommended distribution.

- **Sound Card or Device:** For peer-to-peer data modes such as PSK31, a good quality sound card is necessary. In most cases, the sound card in a modern laptop is sufficient; however some HAMs prefer to use external sound cards to obtain the highest quality with their digital transmissions.

A big advantage of transceivers like the TS-480SAT is that they come with a plug-and-play analog audio connector for use with digital modes. For older transceivers without this connector, you'll need a custom adapter that enables the audio ports on the computer to work through the transceivers PTT microphone and external speaker connectors.

- **Terminal Node Controller:**   The big advantage of modern transceivers like the TS-480SAT is that they come with terminal node connector (TNC) connections which can also be used with multimode communications (MCP) or multiple protocol (MPC) devices, which perform the same roles as a TNC plus more.

Therefore, for full connectivity, your HF transceiver requires connector ports for serial control and data communications with external devices.

Serial Communications Connector

A serial communications port allows a computer to directly control the transceiver via software designed to emulate the front panel controls on the transceiver or to operate an external component such as a data mode terminal node controller (TNC).

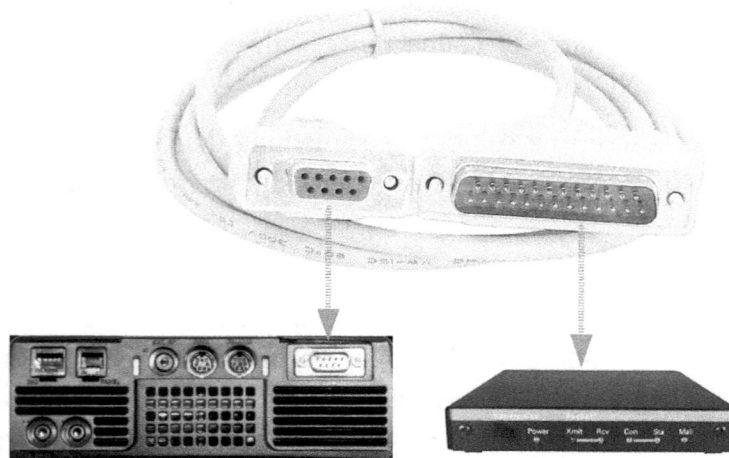

To locate the serial communications connector port on the back of a TS-480SAT transceiver, look for a 9 pin, DB-9 male connector. This connector mates with a nine pin DB-9 female connector and cable. On the other end of the cable will be another connector, DB-9 or DB-25, which will mate with a with a serial communication (COM) port on the computer.

Because DB-9 and DB-25 date back to the very first days of dial up modems, these connectors are rarely seen on a modern desktop or laptop computers these days. For this reason, USB communication port adapters may be required unless your device comes USB ready. With these adapters, the serial cable uses a USB com port adapter with a standard USB 3.0/2.0 connector to complete the connection path to the computer.

USB to COM port adapters normally work well with Windows computers straight out of the box. However, with Linux distributions, the adapter must support Linux (not all do) and some programming at the administrative level to assign a COM port is required. For Linux distributions, you'll need the help of the software engineer on your technical team.

With other transceivers such as the Kenwood TM-V71A VHF/UHF the serial communications port connector will be an 8-pin female mini DIN port. In this case, a special cable is provided by the manufacturer with an 8-pin male mini DIN connector for the radio connection and a 9 pin, DB-9 male connector for the serial communications connection with your computer.

> *REMEMBER: When evaluating a transceiver for base station use, look on the back panel for an 8-pin female mini DIN port or a 9 pin, DB-9 male connector. You will know the transceiver is ready for serial communications with a computer right out of the box.*

Data Communications Connector

Where the role of the serial communications connector is to facilitate control of an external device connected to your transceiver, the data communications connector is where traffic signals to and from the transceiver are handled. To do this, you'll want a data communications connector on your transceiver as a convenient single connection point.

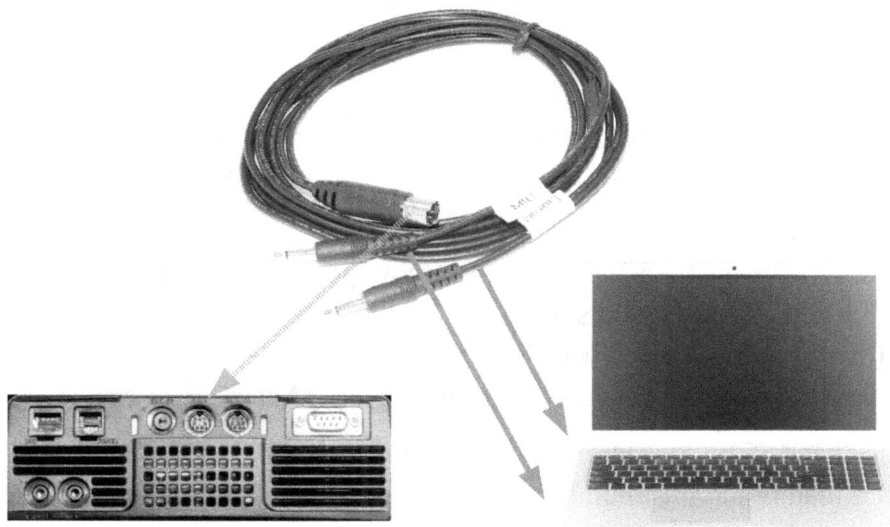

Both the Kenwood TS-480SAT and the Kenwood TM-V71A use an industry standard transceiver data port connection. To locate the data port connection, look on the transceiver back panel for a 6-pin female mini DIN port.

When unpacking your transceiver for the first time, be sure to look for the data port adapter cable. Modern radios will typically have a 6-pin male mini DIN port and two 3.5 mm male audio jacks on the other side. One is for microphone in and one is for audio out on your computer.

In normal operation, your computer will essentially emulate the PTT microphone and speaker on your transceiver. Messages are composed on the computer with software and the computer's sound card which then modulates these binary digital signals into analog signals (tones) your transceiver can transmit. Conversely, signals received as analog signals are relayed to the computer which converts them to digital data for software manipulation.

> *REMEMBER: When evaluating a transceiver for base station use, look on the back panel for a 6-pin female mini DIN port connector. That way, you will know the transceiver is ready for data communications right out of the box.*

Key and Keyer Ports

Morse code in the CW mode with HF transceivers will be an essential part of your community communications strategy. For CW, you will configure your HF transceiver with a telegraph style straight key or an iambic electronic keyer with Morse code automation.

The difference between the two is that a straight key is as simple as a momentary on-off switch. You could literally use two bare wires to make a simple straight key.

Keyers on the other hand require a special hardware interface to handle the necessary automation. Both types will be useful and necessary. In fact besides AM/FM/SSB, the only other mode you will use for your community communications strategy is CW, and these will be your essential modes.

Your beginners will train with a straight key, and when they become proficient, they will begin working CW contacts with HF transceivers. Then, an electronic keyer will help to increase their speed while reducing the risk of "glass arm," (an old CW term for carpal tunnel syndrome.)

If your transceiver is inexpensive, such as the low-power level 1 QRP CW transceivers described earlier, it is likely to have a simple key port but no keyer port, unless you purchase an add-on circuit card for the transceiver. You may also purchase an external keyer device that goes between your transceiver and your electronic keyer. (Again, more cost and complexity.)

On the other hand, CW is where a good HF transceiver like the TS-480SAT truly shines. Not only does it feature a robust number of signal processing features to clean up a distant CW signal, it also offers full support for both keys and keyers straight out of the box.

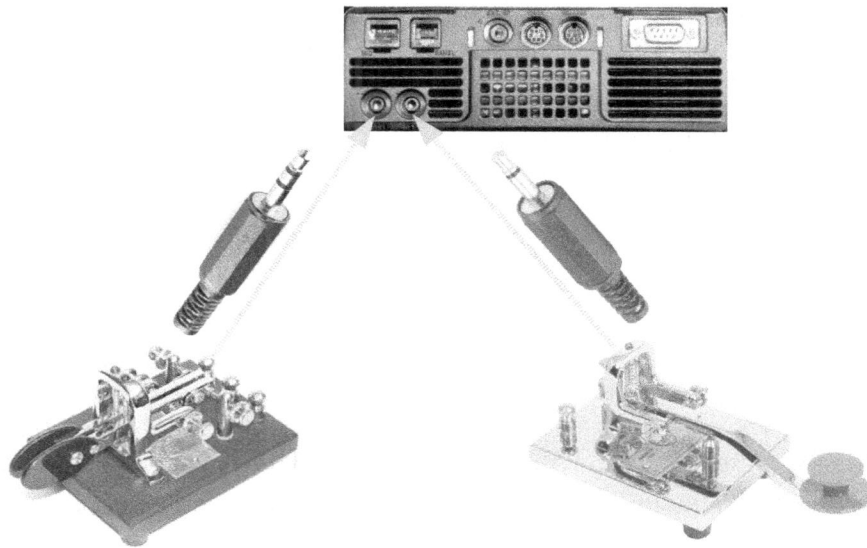

How can you tell if a transceiver offers both key and keyer support? Look at the port where the key or keyer is connected. If the lettering says "Key" and there is no other port, the transceiver is only intended for use with a straight key. However, if you see a second port with the lettering "Keyer," then both types are supported.

Another way to make this determination is the connector itself. Older base station transceivers typically do not have a built-in keyer and use a quarter inch mono plug to connect the straight key to the transceiver.

Because the TS-480SAT supports both straight key and electronic keyers, it uses the prevailing industry standard based on two connector types. For straight key you'll see a 3.5mm female mono connector on the back panel. For electronic keyers, a 3.5mm female stereo connector is used. A nice advantage of this arrangement is that the operator can seamlessly switch back and forth between a straight key and an electronic keyer at will.

Choosing Level 3 HF Options

When compared with channelized transceivers, amateur transceivers are many times more robust, both in terms of features and performance. They offer more transmit power and a wide range of non-phone modes such as CW, RTTY, Data and Image. However, these non-phone modes will require external components.

The most simple will be Morse code keys and keyers, but the other non-phone modes, RTTY, Data and Image, will require a more sophisticated range of components, which in turn will need to be powered and maintained.

For these reasons, at the outset of the global tribulation, you and your technical team will be primarily focused on level 1 and level 2 of your community communications strategy, whereas with level 3, you will have a much wider range of options to consider and choosing the best ones will vary from mission to mission. Here is where you will want to work closely with licensed HAMs who already possess the equipment and skills. They'll know how to do it and with whom to connect, if that is still possible.

In the meantime, simple is the best start and returning to our pizza analogy, it is easy to see that you can eat one slice in a single setting with SSB CB, whereas eating the whole HF pie will take a little longer, but look at all the extra pizza you get.

The point is your level 1 and level 2 HF transceivers will be the fast track entry point to the world of HF for low and high risk missions. Regardless of the transceivers you choose for level 1 and level 2, your level 3 command base stations must fully support them.

Once all this has been accomplished you can expand into other non-phone modes as your needs and abilities dictate. But once you get your level 3 HF transceivers up and running with multiple external components for non-phone modes, the world will be your oyster as they say.

Simply put, amateur VHF/UHF will get you through the end of this civilization and amateur HF will offer a clean slate for laying a foundation for the next civilization. So, given all that is at stake, what is a good way to determine which non-phone modes you need to support?

The smart way is to implement a signals intelligence gathering plan along with your community communications strategy. You will then have an inexpensive and effective way to learn about the world far beyond your local operating ranges. You can learn about who is talking, what they are talking about, and whether you want to talk to them.

16

Signals Intelligence (SIGINT)

The ability to hear what is happening in the world around you is just as important as being heard, if not more so. Therefore, the goal is to gather radio frequency traffic in your area, so as to passively gather what is called signals intelligence (SIGINT).

The risk of using expensive and difficult to replace transceivers for SIGINT gathering during daylight hours can be mitigated with the use of less expensive receivers, including mobile scanners and shortwave receivers for Long Distance (DX) SIGINT on a 24/7 basis.

You can locate a SIGINT listening station inside your base station, but the best approach will be to keep your SIGINT teams mobile with battery-powered handheld scanners and short-wave receivers. For two-way communication between the SIGINT team and your command base station, an amateur HT transceiver or a backpack mobile will allow the team to quickly move to the best locations for reception.

Also, equipping your listening teams with a good quality spotting scope or binoculars for visual observations and visual signaling devices is also advised, depending on the tactical situation.

Conventional Analog Scanners

Mobile scanners, also known as "police scanners", are used to monitor conventional analog channels and frequencies. The use of conventional analog mobile scanners has declined sharply as municipalities move to more secure trunking and digital radio systems.

While scanners with trunking and digital scanning feature are becoming more available they can cost five times as much as an inexpensive conventional analog handheld scanner, which will be more suitable for your actual signals intelligence. Furthermore, they cannot decode the transmissions so even if you can receive an encrypted digital signal from a municipality, all you will hear is useless noise.

While expensive scanners can receive signals on thousands of frequencies, less expensive handheld scanners typically offer fewer local SIGINT bands and frequencies, such as:

Spectrum	Band	Description	Mode	Frequencies
HF	11 m	CB	AM/SSB	40 Channels
HF	10 m	Amateur HF	AM/FM	28-29.7 MHz
VHF	6 m	Amateur VHF	FM	50-52 MHz
VHF	2 m	Amateur VHF	FM	144–148 MHz
VHF	2 m	Business Band	FM	150-156 MHz
UHF	70 cm	Amateur UHF	FM	430-450 MHz
FRS/GMRS	65 cm	Consumer UHF	FM	462-467 MHz

Scanning other services, bands and frequencies aside from these will depend on the receive range of the scanner. Obviously, the more bands and frequencies the better, but keeping to basics is just as important. When evaluating conventional analog scanners, the basic criteria worth keeping in mind are:

- **Retail Cost:** Your teams will use scanners on a 24/7 basis which means daytime exposure to solar storms is a factor. Because redundancy is important, scanners costing under $125 retail are preferable to more expensive scanners.

- **Fixed and Mobile:** The scanner must work equally well in a fixed location as well as on the move. Not all scanners are equally capable of both so do your homework.

- **Type:** Scanners are available in both handheld and base station/mobile versions and in some cases for comparable prices. While you will likely want to use mobile scanners in your command base station, handheld scanners will be more suitable to listening teams operating above ground.

- **Power:** For handheld analog scanners, dry cell and rechargeable AA batteries are preferable. If specialty batteries are required, use a high level of redundancy. You may want to consider solar rechargeable battery systems.

- **External Antenna:** Use of high gain aftermarket antennas with a BNC or SMA mount can greatly enhance the reception of a scanner. Whatever handheld scanner you chose, make sure that its antenna is removable and not fixed.

- **Audio Output:** The ability to record signals with a handheld voice recorder is important. Make sure the scanner has a headset or line out jack.

A good reference example of a suitable handheld conventional analog scanner for your listening teams is the Uniden BC125AT 500 Channel Handheld Scanner with Alpha Tagging. Available online at just under $100, it is packed with useful features and can be programmed using a computer.

If you want to install scanners in your command base station, an inexpensive mobile scanner is the Uniden BC355N 800MHz Base / Mobile Scanner. Costing not much more than a BC125AT, it includes all the same frequencies plus 800 MHz public safety frequencies.

Above the BC125AT and the BC355N, the flagship of the Uniden line is the BCD536HP Digital Phase 2 Base/Mobile Scanner with HPDB and Wi-Fi.

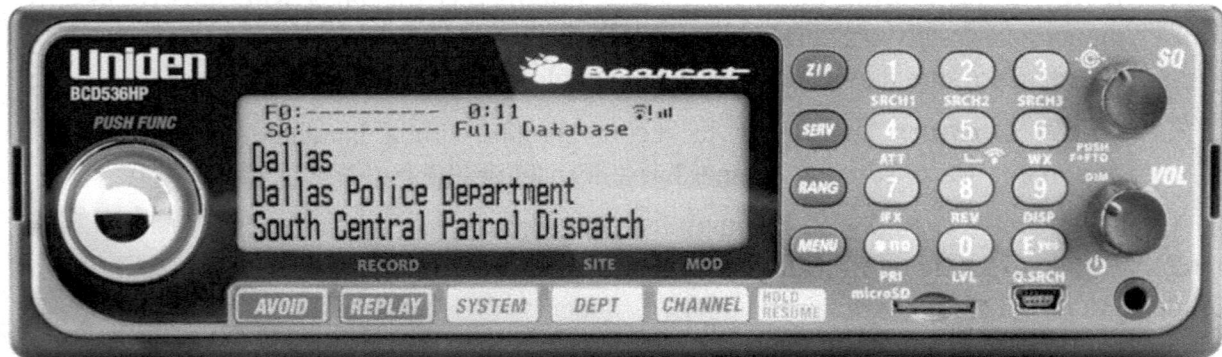

A remarkable scanner that costs five times that of a BC125AT handheld scanner. Or in other words, for the price of one flagship scanner, you can acquire a less capable scanner in much greater quantities for the same amount.

Also note, while Uniden has a broad range of quality scanners, there are other manufacturers of quality scanners, such as Whistler, which should also be considered by your technical team.

Further to that, your technical team should also evaluate mobile shortwave receivers for HF DX monitoring as well scanners.

Shortwave Receivers

Most conventional analog scanners will include the 10 m amateur band in addition to 11m CB and the 6m VHF "magic band." The best will scan all HF bands in both AM and SSB mode as well as commercial AM and FM station broadcasts.

For long distance (DX) SIGINT a handheld shortwave receiver should have a built-in antenna plus a port for an external wire antenna. There are several popular brands to choose from, including Eton, Sony, Sangean and Tecsun.

While the available features vary widely between brands and models, the key criteria for your listening post teams will be durability, ergonomics and usability.

When conducting your evaluations, one shortwave receiver that sets the bar for cost and features is the Sangean ATS-909X. In terms of cost, it is in the middle of the pack with a typical retail price of about $200.

The ATS-909X is the flagship of the Sangean line. It offers a wide range of bands and modes including single side band (SSB) with both upper side band (USB) and lower side band (LSB) support. For listening teams, the most useful features of this radio are:

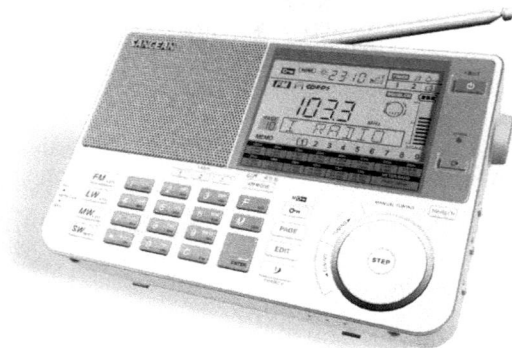

- **Excellent Ergonomics:** The display and controls of the ATS-909X are some of its best virtues. Unlike other transceivers with tricky tuning buttons, flimsy built-in antennas and small displays, the ATS-909X offers a reliable benchmark for comparing other brands and models.

- **Multiple Tuning Methods:** Scanners like the ATS-909X offer excellent scanning capability with five tuning methods: direct frequency tuning, auto scan, manual tuning, memory recall and rotary tuning.

- **External Antenna:** Sangean makes an external wire antenna, the ANT-60 SW which is designed to work with the ATS-909X. Buying one without the other is pointless.

- **PLL Synthesized Receiver:** A common feature with modern shortwave receivers is the use of phase-locked loop (PLL) tuners. Well-designed receivers with (PLL) tuners can improve signal reception and can recover signals from a "noisy" communications channel without annoying birdies. (A birdie is a false or phantom signal caused internally by the circuit design.)

At this point, we've introduced several types of scanners and receivers for SIGINT gathering and the key point here is that scanners and receivers cost much less than transceivers and generally do a better job of scanning local traffic. This makes them affordably expendable, especially during daytime operations.

Here the operative term is "daytime" because this is when your electronics will most vulnerable to the threats of lightning strikes and Solar (CME) power surges. This threat must not be underestimated because during a global tribulation, the frequency and severity of these events will become less predictable and more extreme.

For this reason, the next part of the book Lightning, CME and Electromagnetic Pulse (EMP) offers useful tips and insights into mitigating these threats.

Part 5 – EMP and Lightning Defense

17

Voltage Spike Threats

Previously we've discussed the power specifications of the different transceivers in terms of volts, amps and watts. However, when delving into the crux of voltage spike threats to include lightning strikes and electromagnetic pulse (EMP) threats, the term "current" best defines the real issue of concern.

Current is a flow of electric charge, resulting from the flow of electrons through a wire. Therefore, if a bolt of lightning directly strikes your antenna, it will follow a current path through the wire connections leading to ground through your transceiver and destroy it.

With knowledgeable HAMs, the catastrophic consequences of a direct lightning strike are a given whenever an antenna that is connected to a transceiver takes a hit.

For those who are new to amateur radio, there can be confusion about this because of how we protect our homes and structures from direct lightning strikes with rooftop lightning rods.

For example, direct lightning hits are a huge threat for those living in Florida, and they know full well that you don't get a second chance with lightning. It is why they say "when thunder roars, go indoors." In most cases, even being in an unprotected structure is preferable to being exposed to direct lightning strikes.

This is because when a home takes a direct lightning strike, the current will tend to follow the available paths to ground, including the electrical wiring, plumbing, cable or telephone lines, antennas and/or steel framework. Therefore, to protect their homes from direct lightning strikes, prudent homeowners will install a series of lightning rods on their rooftops.

These lightning rods are connected by a common length of lightning cable which leads down to one or more grounding rods installed near the home's foundation. The aim is to keep a direct lightning hit from going into the home and by diverting it to ground before it can enter through the home's electrical junction box, telephone wires, etc.

A critical point to keep in mind with lightning rods is that there are no inline components or electrical devices. Rather, they offer a simple path. The current of a direct strike follows a simple path. Lightning rod → to lightning cable → to grounding rod → to soil.

Added protection from indirect hits is also addressed with a home lightning arrester and other voltage spike protection devices. But what happens if a direct lightning bolt (which is five times hotter than the surface of the sun) mysteriously ignores the home's rooftop lightning rods and strikes the power line near the home lightning arrester installed on the home's electrical junction box instead? In this case, the arrester along with the wiring in the home and many, but not necessarily all connected electrical devices, including every type of surge arrester or protector that is connected with the current path and much of the wiring, can or will be severely damaged or destroyed. Lightning is a capricious thing and can sometimes choose a path to ground that ignores connected electrical equipment.

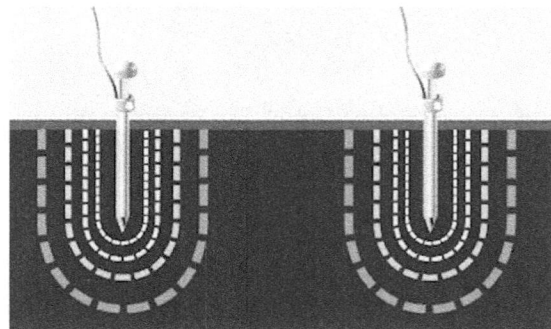

This is why the mention of direct lightning strikes is often buried in the marketing literature if it is mentioned at all. After all, who wants to buy an expensive protection device if a direct lightning strike will turn it into smoldering bit of wreckage without protecting anything?

A less-often considered type of damage occurs when a nearby lightning strike causes damage due to electrical currents induced by lightning's magnetic field. This can occur even where there is no electrical connection.

Another confusing aspect in the marketing literature is how these power protection devices are named. They are frequently labeled as lightning / EMP surge protectors, even though lightning strikes are not actually true electromagnetic pulse (EMP) events.

EMP Events

What is a true EMP event? A true EMP results from events such as a coronal mass ejection (CME) solar storm event or an atomic bomb blast. Ergo, if lightning strikes are not true EMP events, then what is the real difference between power spikes created by lightning strikes and true EMP events? The answer – speed and field strength.

The speed of a lightning strike is typically measured in increments of microseconds, whereas the speed of an EMP event is typically measured in increments of nanoseconds, or 1000 times faster.

For this reason, indirect lightning strikes are often survivable events because they require time to build current, which being significantly less than a direct strike gives power spike protection components just enough time to respond. Therefore, indirect lightning strikes may be survivable with adequate protection, but direct lightning strikes are not because they overpower the protection devices with millions of amps.

Even though the speed of direct and indirect lighting is typically measured in increments of microseconds, a direct lighting strike will hit with a massive and overwhelming current strength that overcomes every component in the current path and destroys it.

In a manner of speaking, one could say that a direct lightning bolt strike is much like getting hit by an electrical freight train traveling at the speed of light. Everything connected to the current path will be destroyed.

While field strength as a catastrophic factor is relatively easy to imagine, the speed difference between events measured in microseconds (one millionth of a second) vs. nanoseconds (one thousand-millionth of a second) is more difficult to imagine.

To help illustrate this speed difference between lightning and an EMP event we'll use an example where a local dam has suddenly failed without warning.

For the people living downstream of the damn, first they hear the noise from the dam's failure and then come the water and debris. It does not come all at once, but rather, the flooding builds in intensity as the dam's reservoir empties.

This is analogous to an indirect lightning strike and those living far downstream will have a brief window of opportunity to get out of harm's way.

The point here is that the speed (measured in microseconds) will give properly configured power spike protection equipment enough time to halt or divert the current from an indirect lightning strike before it can damage your equipment.

But what about those folks living just below the dam? They are not as lucky because the flood waters will come upon them so quickly; they'll be smashed to death by the flood waters before they can react. This is analogous to the speed of an EMP event (measured in nanoseconds) such as a nuclear bomb blast. This is not to say that power spike protection is a waste of time and money.

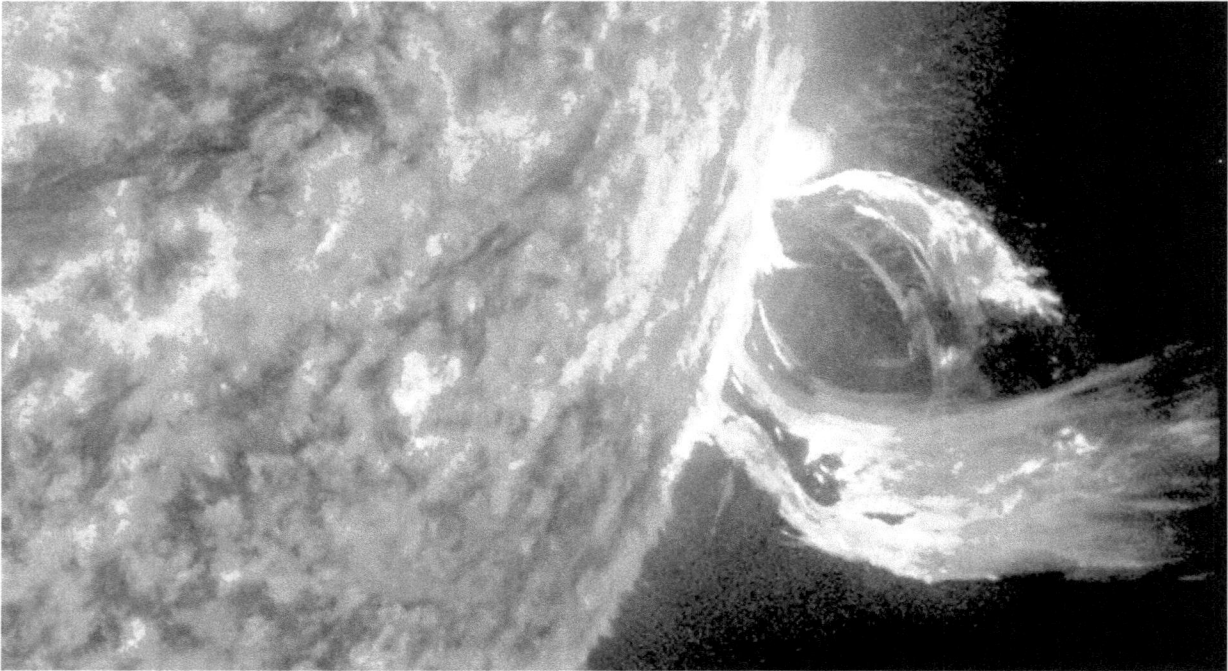

A typical CME, while generating massive magnetic fields and therefore electrical currents, occurs over a much longer period of time, seconds to minutes. Therefore any device that is connected to an appreciable length of wire will likely be destroyed by a CME.

Playing the Percentages

While there is no absolute way to protect electronic components, it makes good sense to do what you can to mitigate the threat. In essence, you play the percentages. To help illustrate the point, let's see why lightning remains a constant concern for radio operators and those who maintain mountain top repeater station systems.

With our example, there is a mountain top repeater station with five different antennas mounted on a tall antenna mast and a lightning bolt makes a direct hit on the third antenna down on the mast. This means one antenna sustains a direct hit and four sustain indirect hits.

The result is that when technicians arrive to assess the damage, they find the transceiver that was connected to the antenna which sustained the direct hit. What they find is, a smoldering wreck, literally blown out of the rack – on the opposite side of the equipment room, a total loss.

Yet, the other four transceivers in the same equipment rack which were connected to the other antennas are still functioning perfectly, because their power spike protection equipment successfully protected them from being damaged.

This is a very realistic scenario because during the coming global tribulation, lightning will become a frequent and serious threat. This is why it makes sense to invest time and money into power spike protection devices.

There is also another benefit. Not all natural EMP threats are catastrophic. With this in mind, let's take a closer look at EMP threats and Cold War survival strategies in the context of the coming global tribulation.

Cold War EMP Threats

There are several types of EMP. Some are natural events like CMEs and others are man-made, such as when a nuclear weapon is detonated. The detonation of the bomb releases an immediate burst of electromagnetic radiation known as a nuclear electromagnetic pulse (NEMP.)

There are also nuclear weapons specifically designed as EMP weapons. They produce a high altitude electromagnetic pulse (HEMP) which is different from the secondary effect of a nuclear EMP. Because they create a gamma radiation flux that strikes the earth's magnetic field in the ionosphere, HEMP bombs generate power spike events with very high field strengths that travel at nanoseconds speeds causing explosive damage for thousands of miles.

Interestingly enough, older analog components are far less susceptible to damage from HEMP than modern digital components, which could help explain the present day EMP definitions which broadly include lightning. An example of this is the 1962 Cuban Missile Crisis.

EMP in 1962 vs. Today

In 1962, the whole world dreaded the prospect of a full nuclear exchange between the United States and Russia. It was a frightful time for everyone on the planet. Had a limited exchange occurred, what would have been the results in 1962 as compared with today?

During the Cuban Missile Crisis, we lived in a world of analog components such as vacuum tube radios and early auto ignition systems, which are more resistant to destructive bursts of EMP radiation. Also, the nation's power grid was different then.

Today, we rely upon massive power transformers that only a few countries currently manufacture and which may take years to construct and deploy. Yet, these massive modern transformers are even more vulnerable to EMP events than Cold War era power grid transformers, which were smaller and more easily and rapidly replaced.

Therefore, when we compare the components and power grids of 1962, we see that we are far more susceptible to EMP today than ever before and this raises a question. How resistant were early Cold War era analog components to EMP?

The answer came in the 1980s with the defections of Eastern-block pilots. A time when our own Air Force was increasing the use of digital avionics, targeting systems, etc. in their aircraft.

When Air Force experts examined the Soviet fighters and fighter bombers flown to the West by the defectors, they were surprised. Even as USA was going digital, the Soviets had gone in a different direction. They had developed next generation, analog vacuum tube avionics.

What American scientists saw was that the Soviets had created a whole new generation of tiny, miniaturized vacuum tubes to replace older and bulkier designs and that these ingenious next generation vacuum tubes were well shielded and virtually impervious to EMP radiation.

At the time, our Air Force employed early digital avionics which were hardened against EMP radiation, though they were still vulnerable to a small degree. However, the need to drop a bomb on a target with precision was the deciding factor.

Hence, after dropping their bomb, American pilots faced a possible loss of control from EMP radiation (though our Air Force eventually closed the EMP survivability gap.) The Soviets on the other hand, had chosen total survivability over early digital sophistication.

With this in mind, what about the modern components that define our present world including the radios discussed in this book? How vulnerable are they to EMP?

The EMP Commission, first created by Congress with the Floyd D. Spence National Defense Authorization Act for Fiscal Year 2001, answered that question with a very real EMP attack scenario.

According to the EMP Commission, the nation's power grid is presently vulnerable to catastrophic EMP-induced failure. This vulnerability is unnecessary because the grid can be protected using existing technology. Furthermore, the cost of such a program was estimated at over three billion dollars. To put that sum in perspective, a modern aircraft carrier costs approximately 12 billion dollars or more to build.

Although the commission found that only ten percent of Americans are expected to survive a catastrophic collapse of the national power grid, opposition to EMP protection by the power industry is fierce and unrelenting.

Power companies are seeking to prevent any legislative incursion into their self-regulated industry. The result is that their K Street lobbying efforts have effectively blocked any attempt by Congress to protect the nation's power grid against EMP threats. The result is that energy companies are willingly putting a death sentence on any pharma-dependent child or adult, such as diabetics and those requiring kidney dialysis.

Modern HEMP Attack Scenario

In 2009 a post-apocalyptic and very successful science-based-on-fact novel titled *One Second After* by William R. Forstchen and William D. Sanders was published.

The plot of this fiction title is based on the EMP Commission findings, and begins with an EMP attack on the United States by two rogue states. Working together, our enemies launch two medium-range ballistic missiles over the USA, each armed with a powerful, HEMP (high altitude EMP) nuclear warhead. These two strategically placed detonations covered the entire nation and parts of Canada, Mexico and Alaska.

The instantaneous result of the attack was that every digital, microprocessor-based component, including those in cars, computers and medical equipment and so forth were destroyed. The result was an immediate collapse of our economy. Death tolls skyrocketed, even though there had been no physical destruction on the ground from these high-altitude attacks.

In a poignant moment in the book, early on a town physician predicts with stunning accuracy who will die and when, by referring to patient's medical records. The doctor's predictions are based on those dependent on modern pharmaceuticals, like diabetics are, to stay alive. These would go straight to the top of the casualty list ahead of the expected causes of violence, starvation, and so forth.

However, at the end of the book the effective impact of two HEMP weapons becomes sickeningly apparent as the main protagonist learns from an Army officer that nine in ten Americans have died. Nine in ten – preposterous you say?

Consider this. The scenario in *One Second After* is so credible, that the book was cited on the floor of Congress and widely read by the nation's leadership. It accurately shows how vulnerable we were to an EMP attack in 2009. 90% of all Americans would die as a result of two HEMP devices delivered by medium-range ballistic missiles and no other attacks.

As for today, have the risks changed? Yes, the risks have changed for the worse, especially when we add a nuclear attack scenario to the overall global tribulation threat matrix, which also includes earthquakes, tsunamis, volcanic eruptions, etc. For this reason, a deeper discussion of natural EMP events is necessary including, Meteoric, CME and Solar Sprite events.

Meteoric and Solar EMP

With a meteoric EMP event, the entry of a sufficiently large meteor into our atmosphere causes ionization and a detonation in the air with essentially the same result as a HEMP weapon. Or, it impacts the ground and causes EMP. Lower levels of measurable EMP can also be caused by volcanic eruptions and Earthquakes, which would likely follow a ground impact event.

During the global tribulation, meteor showers and impact events will be an ever-present concern and if your system is well-protected from indirect lightning strikes, it can survive a small to moderate meteoric EMP event relative to your location.

A Solar EMP is caused by a type of solar storm called a coronal mass ejection (CME). Typical CME solar storms come in two stages. The first is a flare, which is an explosive release of energy that reaches Earth at the speed of light (approximately eight minutes).

These Earth-directed flares carry very high-energy particles, such as those carried by CMEs. They can disrupt communications and in severe cases, can cause radiation poisoning to humans and other mammals.

While a solar storm may not always begin with a flare, there will always be the slower-moving CME, which is the explosive burst of particles and electromagnetic fluctuations that typically takes between 12 hours to a few days to reach the earth.

Our sun on average produces three solar storm events each day; so why do these events rarely affect us? There are three reasons: direction, strength, and magnetic orientation.

When a solar storm erupts, the first concern is direction. If the direction of the storm takes it away from Earth and into space, it is not a concern, which is the case for the vast majority of solar storms.

However, when the direction of a solar storm puts our blue-green planet in the cross-hairs so to speak, it is what is known as an Earth-directed CME and these are the worrisome type, when they are large and strong.

To assess the strength of the solar flare, the flare that precedes the CME plasma is rated on a scale with the following classifications: A, B, C, M, X and Y.

- **A, B and C Class:** The smallest flares, they are too weak to noticeably affect the Earth.

- **M Class:** When Earth-directed, these can cause brief radio blackouts near the poles, but will not damage your radios.

- **X Class:** When Earth-directed, X class events are large explosions and loops on the sun can be tens of times larger than the Earth. These events can create long lasting radiation storms that can harm satellites, ground-based components and power grids.

- **Y Class:** This is a whole category added to the scale in 2013, when the strength of a massive solar storm jumped off the X class scale. Y class events are extremely rare at present, but will become far more prevalent during the global tribulation and when these are Earth directed, they will do far more damage than an X class event.

The reason why a Y class event will do far more damage than an X class event is due to how the scale is calculated. Each letter represents a 10-fold increase in energy output. Ergo, a Y is ten times that of an X, and 1000 times that of a C.

When an Earth-directed X or Y class event occurs, the magnetic orientation of the CME becomes the real determining factor. Astronomers call this the z component and it describes the north-south direction of the CME's magnetic field.

To be catastrophic a CME must be Earth-directed and its z component must be negative, which corresponds to a southward magnetic field. As such, it will disrupt satellites and penetrate Earth's atmosphere, inducing electric fluctuations at ground level. The result will be blackouts as power grid transformers fail turning modern digital components into useless junk.

On the other hand, an Earth-directed X or Y class storm will be more survivable provided it a positive z component with a northward magnetic field. In this case, if you have a robust power spike protection system, the odds are your components will survive.

So is an Earth-directed Y class storm with a southward magnetic field the worst case scenario? No. The worst case scenarios are reserved for something called a solar sprite.

Solar Sprite EMP

Solar sprites are also known as cosmic lightning because they start with a sun-based lightning that arcs across space and smashes into a planet with catastrophic results. At present, solar sprites emanating from our sun have not been observed in our solar system. These events can be argued as theoretical but they remain undeniable.

During the global tribulation, our sun will interact with the objects in this global system, which will likely result in solar sprites. Should an Earth-directed solar sprite strike our planet, it could gouge out an impact scar the size of the Grand Canyon or larger, bringing with it an EMP event of unimaginable proportions.

Should such an event happen, your best hope is that your community is on the night side of our planet when it hits the day side, because wherever it hits, whoever and whatever is on the surface will be vaporized.

It is for this reason, the elites have constructed their survival shelters called Deep Underground Military Bases (DUMBS) some 2 miles below the Earth's surface. These are the only real protection against a solar sprite.

What is the point of all this?

Accept the fact that there are things you cannot protect your community from. Rather, you play the odds and do everything you can to mitigate a survivable event. For this reason, you want to implement a robust, voltage spike mitigation strategy.

18

Voltage Spike Mitigation Strategy

During the global tribulation, your community will face a broad range of voltage spike threats and there is no absolute form of protection; no silver bullet if you will. For this reason, you need a robust voltage spike mitigation strategy. Options include the following methods:

- Alternative Communications
- Administrative Mitigation
- External EMP Defense

Do not cherry pick a few options from below for your voltage spike protection strategy. Rather, you should throw the proverbial kitchen sink at this – starting with alternative communications.

Alternative Communications

During a global tribulation, there will be no such thing as above ground survival shelters. Solar-driven events, such as those discussed earlier, will affect the Earth's four spheres: atmosphere, biosphere, hydrosphere and lithosphere. This in turn will create violent storms, eruptions, earthquakes, fires, high velocity winds, and other disasters that necessitate the construction of below ground shelters.

Alternate communications, such as landline phones, signal lamps, wired and WiFi intranet, are excellent options for large underground survival communities with a population of one hundred or more members and multiple structures. For smaller groups in a single structure, the cost of alternate communications may be difficult to justify, but they are just as necessary.

Soil Cover

Sufficient soil cover above the shelter can protect components from most forms of voltage spikes provided they are not connected via wire to an above ground conductive source such as a power feed or coaxial antenna line. Any above-ground wired communication system, particularly any with long runs of wire, will be an antenna for voltage spike events. Once a spike enters a shelter, it can create havoc with electrical/electronic equipment.

Thus, burying the wires at a minimum depth of five feet below ground (this author recommends a depth of 10 feet) will significantly reduce the danger of damage and injury or death. For optimal protection, above-ground wired connections should be disconnected and floated during extreme weather events and daylight hours.

Below Ground Intranet

A below ground solution for large communities with multiple below ground structures with one hundred survivors or more, will be an underground intranet, like those used in offices across the country for sharing resources.

A laptop or personal computer is configured as a web page and file server and is connected to the intranet with a router (like those used in home and office). It should support both hardwired, ideally fiber-optic connections for other computers, and Wi-Fi for smartphones and tablets. The advantages here are modest power demands, above ground antennas are not required, and this reduces the use of valuable transceivers. Another alternate communication device for daylight use is a signal lamp.

Morse Lamp

For 24/7 Morse code signaling above ground, a highly survivable technology is a Morse lamp. Whether you're using a signal lamp or a transceiver, it is essential that Morse code proficiency become a requirement for your community communications strategy.

Signal lamps, also referred to as Aldis or Morse lamps, are simple to operate and date back to late 19th century. Still in use today by navies and airports worldwide, they transmit Morse code messages via optical communication (flashing lights) as opposed to radio communication.

Investing in Morse signal lamps for alternate communications is a wise move, especially for your level 1 – emergency radio and level 2 – field operation radio kits. When you consider how inexpensive a handheld signal lamp or a small flashlight with signaling capability is the value is obvious.

Analog Mechanical Field Phone

Another alternate communication device with a long history of military service is field landline telephone systems. For example, during the 1962 Cuban Missile Crisis, the US Army used mechanical analog landline systems comprised of TA-312/PT field telephones and SB-22/PT field switchboards.

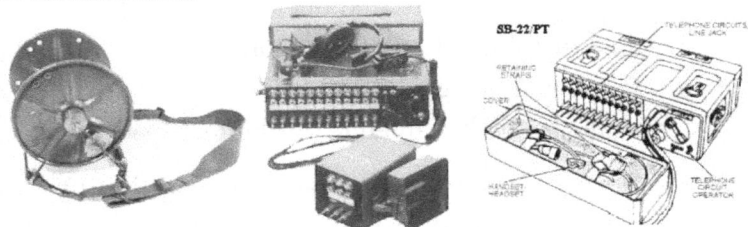

These old mechanical analog phone systems are incredibly durable and highly resistant to voltage spike threats. Use them to connect your above ground community observation posts with your command base station and underground for a simple phone network that connects multiple survival shelters.

Longer distances such as long lines to other survival communities are also possible and here the durability of the communication wire is important. With internal use within your community a 22 gauge wire can carry your signal for up to 15 miles without amplification. However, if you need longer line connections with other communities, use the Mil Spec 18 gauge copper-clad steel, twisted-pair phone wire instead. It can give you up to 25 miles of range.

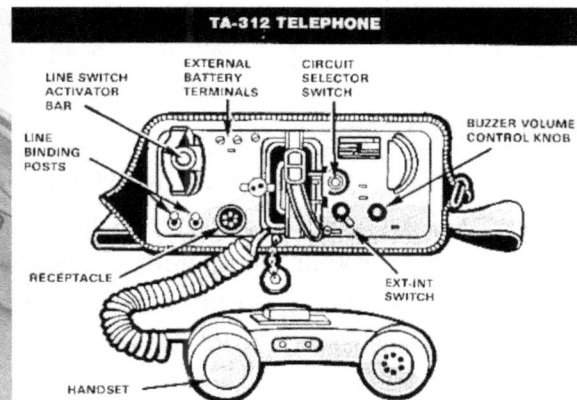

EMP Proof Phone System
Korea / Early Vietnam

TA-312/PT Field Telephones
SB-22/PT Field Switchboard

There are other analog phone system options as well and regardless what you choose, analog landline phones all have one important benefit, low power drain, and many will operate quite effectively with common flashlight batteries or earth batteries.

The benefit of augmenting your two-way radio communications with analog landline phones is that you will reduce the exposure of sensitive components to voltage spikes. Be absolutely sure to keep any wired landlines and components well away from other electrical/electronic systems, especially your valuable radio communications and computers.

While these are effective physical solutions for mitigating the risk of damage to vulnerable electronics from lightning and EMP voltage spikes, there are also effective administrative options as well.

Administrative Mitigation

Of the different voltage spike protection strategy options, administrative mitigation will eventually prove to be the most effective over time and there are just three basic options.

Redundancy

Redundancy is the provision of additional or duplicate systems and this option has been discussed several times, because redundancy is the most effective way for communities to maintain a viable community communications strategy. Minimum redundancy is a factor of three with one operational component and two spares.

However, a redundancy factor of six or more will be far more likely to get a community through a decade long global tribulation.

Of all the various strategies for ensuring your ability to maintain your communication operations throughout a global tribulation and beyond, redundancy is your best strategy.

Scheduled Use

The most risky time to operate two way radios will be during daylight hours since this is when EMP spikes occur from solar storms and CME events. Here is where you can implement a "No Glow" policy for your community. Before connecting a two-way radio to an above ground antenna, make sure that it is only done during nighttime hours. This means that you will stop operating the transceiver before the first glow of sunrise and will not begin operation until after the last glow of sunset.

Regarding meteoric EMP and nuclear EMP weapons, these can occur any time. Therefore, it is advisable that the only time you connect antennas and power up the two-way radio is when you need to transmit a signal. Remember, never power up and operate any transceiver until you have connected a suitable antenna (or dummy load.)

Transmitting without a proper antenna can destroy a transceiver, but in many cases, the transceiver will be damaged but repairable. The at-risk component is called a final RF Power output transistor. These transistors are relatively inexpensive if they are available for current models. For older and obsolete models, they may be difficult to find. It is advisable that your technical teams stockpile them to repair salvageable transceivers.

When you have field teams operating beyond your outer perimeter, there will be times when you have to take the risk of continuous operation for field team transceivers and the repeaters servicing them as opposed to scheduled use. Here is where you simply have no other option than to take a calculated risk and hope for the best.

Solar and Weather Forecasting

One of the most important things you can do as the leader of a survival community and the founder of a local Radio Free Earth Network is to create a forecasting group within your community of individuals with experience in the field of meteorology. These team members will not only look for local weather issues that will affect your operations and those of your listeners, but looming solar threats as well.

For this reason, it is advisable to include a basic solar observatory in your community communications strategy. There are a range of options which include a good quality telescope with a Coronagraph attachment or you can do something less inexpensive such as projecting an image of the Sun onto an improvised screen using a pinhole camera or a pair of binoculars.

Of the three options, solar and weather forecasting offers a highly valuable way to build news gathering connections with other communities via long distance (DX) communications. Reporting these forecasts with scheduled and emergency broadcasts to your local audience will greatly enhance the perceived value of your community by other communities.

External EMP Defense

With an external EMP defense, the first step is to categorize EMP events. The damage can be mitigated in part or whole with a robust power spike strategy or they can be catastrophic with a certain loss of components.

- **Earth-directed EMP Catastrophic Events:** Strategic and HEMP nuclear weapons, X and Y class solar storms with a southward orientation of the CME magnetic field and solar sprites.

- **Earth-directed CME Mitigation Events:** A through M class solar storms, small to medium meteoric EMP, A, B, C, and M class solar storms. Y class solar storms with a northward orientation, volcanic eruptions and earthquakes.

The strategy here is to keep the Electrical Current away from your components altogether. There are four basic EMP mitigation options:

Earth Shielding

Your base station is located below ground with a minimum of five feet of dirt over the structure. The most effective and affordable Earth shielding is a combination of ten feet of soil or more with a two foot layer of volcanic basalt aggregate rock which provides effective additional electromagnetic shielding.

The goal here is resistance and the advantage of basalt rock is that it is paramagnetic and in terms of CME current, it offers as much resistance, inch for inch, as steel. To save money, purchase basalt tailings instead of aggregate as both are equally effective for this purpose.

Underground Power System

If you are using an above ground generator to power your communication system and other digital electronic, you're only inviting catastrophe.

Never use conductive metallic materials or allow power cables to run to the surface. That will only put a bull's eye on every electronic device powered by that system. That creates a serious problem for backup solar and wind generators. They must be installed with very effective spike mitigation systems and disconnects. Otherwise, they can cause a total destruction of your power systems.

Antenna Floating

Antennas may be grounded or ungrounded and your field teams operating beyond your outer perimeter will typically use ungrounded wire antennas so floating is an excellent option for field operations.

With floating, you disconnect an ungrounded antenna at its feed point and then move the antenna coax cable and connector at least six feet away from any radio gear. Also, never allow the cable connector to touch the ground or to become wet and dirty.

When you do this, an ungrounded antenna will still be conductive, but because it has been floated, the current will have no circuit to complete. Therefore, a power spike will likely move past the floated and ungrounded antenna without damaging it or your equipment.

Above all else, understand and accept the fact that there are no safe absolutes with EMP events that can be potentially mitigated. This is why something is better than nothing and if unprotected components are burned to a crisp that is certainly what you'll have – nothing.

Faraday Cages

The purpose of a Faraday cage is to surround your components with an electrically conductive shield so current is directed around them and not into them.

A lot of folks talk about using microwave ovens for EMP protection, this is more of an urban legend than a workable solution. Granted, a microwave oven will afford some protection, but the best you can hope for with a microwave oven is between 30 to 50% protection.

Instead of using urban legend Faraday cages, you want to use more reliable forms of protection which can take on many forms. The most simple are Faraday bags field operatives can use to shield their field radios when not in use.

Likewise, you can build a large Faraday cage using expanded steel screen or copper mesh with small perforations, that is every bit as effective as solid steel and much less expensive.

Always remember, electrical current like water is invasive and it is always looking for something to equalize it. This means, no matter what type of Faraday cage or bag you use, it must seal completely and be particularly mindful of the corners.

Corners are always an inherent weakness with Faraday cages which is why sealed steel drums and cylinders are often preferable to boxes.

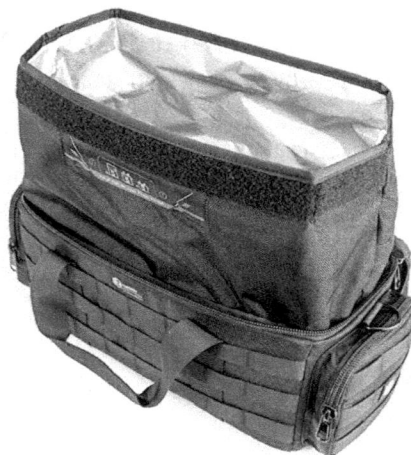

If you are using a box design, sharp angles are to be avoided. For example, a box with rounded corners is much better for protection than one with sharp, right angle corners.

Drum-Inside-a-Drum Faraday Cage

A drum-inside-a-drum Faraday Cage is a relatively inexpensive way to protect redundant stockpiles of valuable electronics. This design uses an inner and outer steel drum configuration.

EMP Protection Using Steel Drums

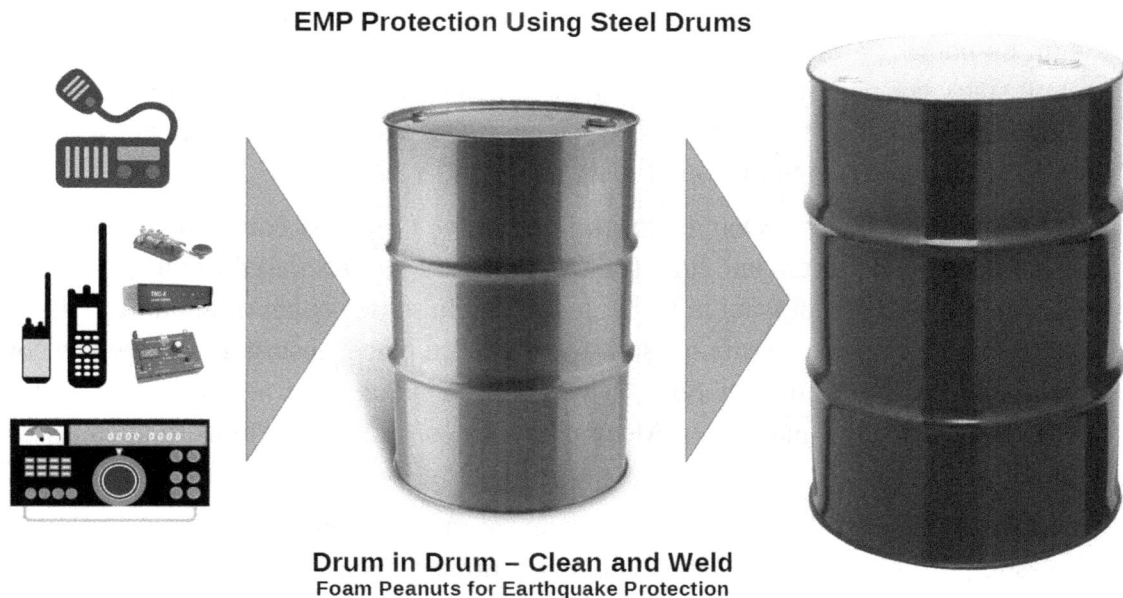

Drum in Drum – Clean and Weld
Foam Peanuts for Earthquake Protection

In the illustration above, you see a 30 gallon steel drum with a removable top, or a galvanized steel 32 gallon trash can, and then a larger, traditional 55 gallon steel drum also with a removable top. Here are the seven steps for drum-inside-a-drum Faraday Cage stockpiling:

1. **Organize Complete Systems:** Prepare kits containing assorted types of electronics that can be opened one-by-one, as needed, during the tribulation. Instead of stockpiling a single type of component, organize your electronics as complete kits with whatever is needed to make them work together as a complete system.

   IMPORTANT: Do not include batteries. While batteries are highly resistant to EMP damage, leakage is a paramount concern. Therefore, batteries need to be stored separately from electronic components to prevent damage from corrosive leakage.

2. **Prepare Both Drums:** Clean and grind off the contact edges between the drums and the lids of both drums, to remove any plastic film or paint. Once the plastic film has been sanded off, you will have a clean, metal-to-metal, conductive contact which is necessary for EMP shielding effect.

3. **Prepare the Electronics:** Before packing your radios disconnect all components and accessories from each other. If the antennas can be unscrewed, remove them. Also remove the power packs, external speakers and microphones, and any other connections. Be sure the device power is switched to the "off" position. Then wrap everything up separately and place it all into stiff cardboard boxes.

4. **Insulate the Inner Drum:** To protect the boxed electronics inside the inner drum, you want to use insulation that will not conduct electricity. The plastic peanuts like those used by shipping companies are ideal.

   Begin by pouring a bed of plastic peanuts on the bottom of the inner steel drum. Then, as you stack your equipment boxes, continue adding plastic peanuts between the boxes and along the walls of the drum. Once you've filled the inner drum with electronics, fill in more plastic peanuts to the top of the inner drum.

5. **Seal the Inner Drum:** Double check that you've created a complete, metal-to-metal contact between the drum and lid. Then weld the lid to the small drum to ensure a tight EMP seal. If you do not want to weld the lid, obtain some copper or aluminum shielding tape from an electronic distributor and tape the lid in place, ensuring that the tape covers bare metal on the drum and lid. Remember, shielding tape adhesive or conductive grease is electrically conductive. Also make sure you store hand tools to open a welded drum.

6. **Insulate the Outer Drum:** With the inner drum sealed and ready, pour another layer of plastic peanuts on the bottom of the larger drum. Then lower the inner drum inside into the outer drum and then surround the smaller drum with peanuts all the way to the top.

7. **Seal the Outer Drum:** As with the inner drum, make sure the outer drum has a complete metal-metal contact between the drum and lid. Then weld the lid to the large drum or as described above, use the shielding tape.

It is also a good idea to make sure your drums are clearly marked. For example, painting them in different colors by kit type will help identify which ones to open first, when needs dictate. This way, if you get hit with an EMP strike that destroys a transceiver, you can quickly pull a complete replacement kit out of storage. Remember, once a drum has been opened unless it is properly resealed, it will be of limited protection to EMP events.

The voltage spike mitigation strategies presented in this chapter offer a sensible and affordable way to protect vulnerable electronics from lightning and EMP. While there are no absolute guarantees, you can also increase your protection with an inline lightning defense strategy.

19

Inline Lightning Defense

Voltage spikes from lightning and EMP present a serious risk for modern day HAMs. Even with the best protection money can buy, a direct lightning strike can still be catastrophic.

In cases of indirect lightning strikes, the damage can be mitigated or blocked as well as a few types of low grade EMP events. Therefore, your inline lightning defense strategy is about protecting radios that are connected and operational on a 24/7 basis.

Therefore, the goal of your inline defense strategy must be to prevent or mitigate voltage spike spikes redirecting current to ground, before it can follow a wired pathway into your connected components and thereby cause damage.

The key to making this work lies in the very nature of voltage spikes.

Current is unpredictable, the one constant with voltage spikes is that they always seek to equalize themselves as quickly as possible. Consequently, voltage spikes naturally seek a path that offers them a way to equalize or neutralize themselves.

Therefore, your goal in creating a pathway for voltage spikes is to makes sure their current is directed to a highly conductive grounding field.

Highly Conductive Grounding Field

This single most important thing you can do when creating an in-line lightning defense strategy is to create a low or no-resistance pathway to a grounding field.

It is why homes built to modern code have one grounding rod or whatever the local building code specifies. However, given the sizable threat matrix of the global tribulation, even that may not be enough.

You should install a robust grounding rod array for your survival community, which is comprised of multiple grounding rods that are bonded (connected) to each other. All grounding wires connected to the ground system should also be connected to the same point to prevent ground loops.

First, you must choose an appropriate site for your grounding field which is very close to your antenna system and radio shack.

> *Be sure that your technical team is familiar with National Electrical Code requirement in Sec. 250-54, which requires the resistance to ground of a single-made electrode (e.g., ground rod) to be 25 ohms or less.*

Soil Survey

A soil survey is necessary, because the effectiveness of your grounding rods will largely depend on the soil surrounding the rods and the degree to which they can conduct large electrical currents.

- **Preferable Grounding Soil:** Virgin moist clay or swamp like soil.
- **Marginal Grounding Soil:** Sandy soils, or dry soils with gravel.

As a general rule of thumb, the best land has rolling hills, soft loamy soil and plenty of water. Sandy soil unless kept wet is not very conductive. When conducting a location reconnaissance for survival community land, use the following table to help determine the best location.

Type of Soil	Soil Resistivity (mega-ohm)
Very moist soil, swamp like	30
Farming soil, loamy, and clay soils	100
Sandy clay soil	150
Moist sandy soil	300
Concrete 1:5	400
Moist gravel	500
Dry sandy soil	1000
Dry gravel	1000
Stony soil	30,000
Rock	10,000,000 (10^7)

Even if the soil around your community is moist, keep in mind that during the global tribulation, deforestation, desertification and loss of surface water will cause the moisture and clay in your soil to change over time.

This is important because even in normal times, this will vary with depth and terrain which also happens to be a constant issue for power companies. Therefore, when acquiring a property for your survival community, you should contact the local power company and ask them to check for grounding rod effectiveness before closing escrow.

Next, let's consider the depth of your grounding rods.

Grounding Rods

Grounding rods come in different lengths from 4' to 10'. Due to the uncertainties of a global tribulation's effects on soil resistivity, here is a simple guideline. For preferable grounding soil 8'-10' copper grounding rods will do. With marginal soil you'll want to use 20' copper grounding rods because electricity has a tough time dissipating into marginal soil. You can also add special electrolyte materials to the soil of a grounding field to make it perform better.

For this reason, resistivity is always the tail that wags this dog, and per the National Electrical Code, the resistivity of the soil for all ground rods must be 25 ohms or less. In contrast to that, the telecommunications industry prefers a lower standard of 5 ohms or less, which is a good goal for your inline lightning strategy. However, given what is coming in the years ahead, it is best to shoot for a soil resistivity of 0 ohms, regardless of the extra cost.

The best way to get there may be with a complex network of bonded (connected) grounding rods. This will increase the amount of contact with the surrounding earth, thereby resulting in lower ground resistances.

Grounding Rod Network

The number of grounding rods in a complex network will depend in part on the size of land available, soil resistivity and design. This is because every grounding rod in your network will have its own sphere of influence.

The ground rod placement formula is simple. If the grounding rod is 10' deep, its sphere of influence will be 10' in diameter. Likewise, if it is 20' deep, its sphere of influence will be 20' in diameter.

If you are using three 10' grounding rods in your network for example, they should be spaced no less than 20' apart. This is very important; because when grounding rods are spaced too close, their spheres of influence will intersect. This can cause problems achieving a lower ground resistance.

Commercial installations will use various types of networks and layouts, but for your in-line lightning strategy, a simple Y shaped grounding array configuration is popular with HAMs because it is effective in dissipating current into the soil. Now let's see how the pieces connect.

The first connection to the ground field will come from a copper entry panel. If copper is not available for the entry panel, brass is the second best choice and aluminum if copper or brass is unavailable. When using aluminum, be vigilant for aluminum oxide resulting from corrosion.

No matter how many connections there are in a grounding field, each must connect to a common single point. Likewise, all components requiring a grounded connection must be connected to the grounding field entry plate nearest the grounding field.

If multiple entry plates are needed, such as one for underground components and another for above ground components, all entry plates must be connected to a common entry plate which is connected directly to the grounding field.

Note: all entry plates should be connected with copper grounding cable (# 8 or larger) and the final connection to the grounding field will be on the grounding rod attached to the left side of the Y. On the right side of the Y, an inline clamp to connect the two grounding feed cables to form the complete Y pattern.

When fastening the copper grounding cables to the grounding rods, always remember that lightning does not like to make a hard turn. If your cable zig zags and bends around with sharp angles, the current will jump out of the array at the point of the hard turn to find another source of ground. Not good.

Always use a straight inline clamp for each connection in the array with the exception of the last grounding rods at the top of the Y pattern where the array ends. Here, you'll use something called an acorn clamp for those final connections. The latest innovation in ground clamps is the inclusion of a corrosion inhibiting grease at the contact points.

Today's grounding fields are typically used with surge protectors and surge arresters. They prevent spikes in the power grid from damaging the components on which we rely in our homes and businesses.

However, during the global tribulation, most of the severe surge spike threats to your components will enter through the above ground antennas connected to your below ground base stations. Current will always follow a continuity path from the antenna and through the coaxial feedlines that eventually connect to your radios.

Hence the term inline, which in this case describes any component that is part of a continuous line of cables and devices, beginning where the current enters the circuit and ending with wherever it neutralizes itself. To protect your components, you'll also use other safety measures such as lightning arresters, which are not to be confused with surge protectors.

Lightning Arresters

The popular terms surge protector, surge arrester, uninterruptible power supplies and surge protector power strips describe power protection devices that are inline with a continuity path. Their purpose is to provide your components with stable voltage and to protect them from power surges. They are not designed to provide your radios and transceivers with stable voltage and protection from surges.

The proper term with antennas is lightning arrester. These devices are used inline with the continuity path from the antenna to your components.

They are designed to provide a continuity path to ground for a high voltage spike before failure via fast response and redirection of the current.

Lightning requires a speedy response and in most cases, common fuses and circuit breakers only protect by opening the circuit much too late to provide protection from lightning.

Gas Gap Diode Arrester

This is why a special type of protection device is used for lightning threats to antennas. It is commonly known as a gas tube, gas gap, or gas gap diode arrester. There are several types of gas gap arresters and the cost varies, but the one constant is speed.

Gas gap arresters are the fastest known inline protection device for redirecting current to ground and a basic arrester will be comprised of the following parts and work in the following order:

Lightning Arrester Part	Description
Coaxial input connector	Attaches to the coaxial cable from the antenna.
Inline gas gap diode	Installed in a weatherproof housing with grounding connector. When a high voltage spike ionizes the gas in a gas gap diode, it effectively redirects the current to the grounding connector.
Grounding Connector/Clamp	This is normally mounted to your grounding field.
Coaxial output connector	In normal operations, RF current passes through the gas gap diode to the output connector without attenuation.

Determining the best lightning arresters for your community will depend on your location, budget, and other factors. However, two features you want to insist on as the leader of your survival community is that your technical team makes sure that your lightning arresters offer multiple strike capability and replacement gas gap diode cartridges are available and easy to access.

The use of two inline lightning arresters is advisable for the best results. One is where the antenna connects to the coaxial feed line cable and the other where the feed line connects to the component. Both should be connected to the grounding field via the feed panel. This affords two opportunities to redirect any current coming from an antenna.

Decoupling Coil

To add a little extra global protection to your inline lightning defense, another device, a decoupling coil, can be added between the arrester and the antenna connection to the component. A decoupling coil will increase the inductive resistance to the output side of the lightning arrester. What this added resistance does is make the low resistance of the continuity path to ground more attractive to the current. And here is the good news: It's cheap to make.

Decoupling coils are simple to fabricate using an appropriate length of cable and a one liter plastic beverage container. Create a length of cable that has connectors at both ends and is long enough to wrap 8 to 15 times and then be taped to the container to keep its shape.

Now you're ready to assemble your complete system using the appropriate cables and connectors.

Cables and Connectors

A transmit power strategy was presented at the outset of this book. While the FCC allows transmit power up to 1,500 watts for specific frequencies with general class license or higher, a 100 watt transmit power strategy is recommended for three fundamental reasons:

- **Signal Control:** If you implement an FM broadcasting system when you begin a local Radio Free Earth network, you will be transmitting 1,500 W signals and even higher. However, using a 1,500 W signal for community operations is overkill. It will travel far beyond the outer perimeter and attract unwanted attention downrange. Remember, SIGINT is a two-edged blade.

- **Power Consumption:** What's the difference between a 100 W transmitter and a 200 W transmitter? You need twice as much power for the 200 W transmitters. This will give you clearer signals but not twice as much as 100 W. In fact, that additional 100 W will only increase the received signal strength by roughly 10% which is a marginal trade-off, especially when you consider the added cost and complexity of 200 W operations.

- **Cost and Complexity:** When it comes to cost and complexity, a 100 W transmit power strategy offers the most optimal price-performance and the widest selection of parts and components. Once you go above 100 W of transmit power, the number of options and availability of parts and components drops quickly as costs escalate sharply.

With this in mind, let's examine a short list of cables and connectors to be used or to use with base station installations based on transmit power. (Note, this is a generally-reliable guideline for use with matched antennas with an SWR of 2.0 or less.) (Note: performance and suitability will vary between cable and connector manufacturers and products.)

Some of the connector types listed below are also used on cables that are not listed here. Heliax is such cable. It is a very expensive low loss cable that is commonly used for high power Broadcast applications and Cable TV.

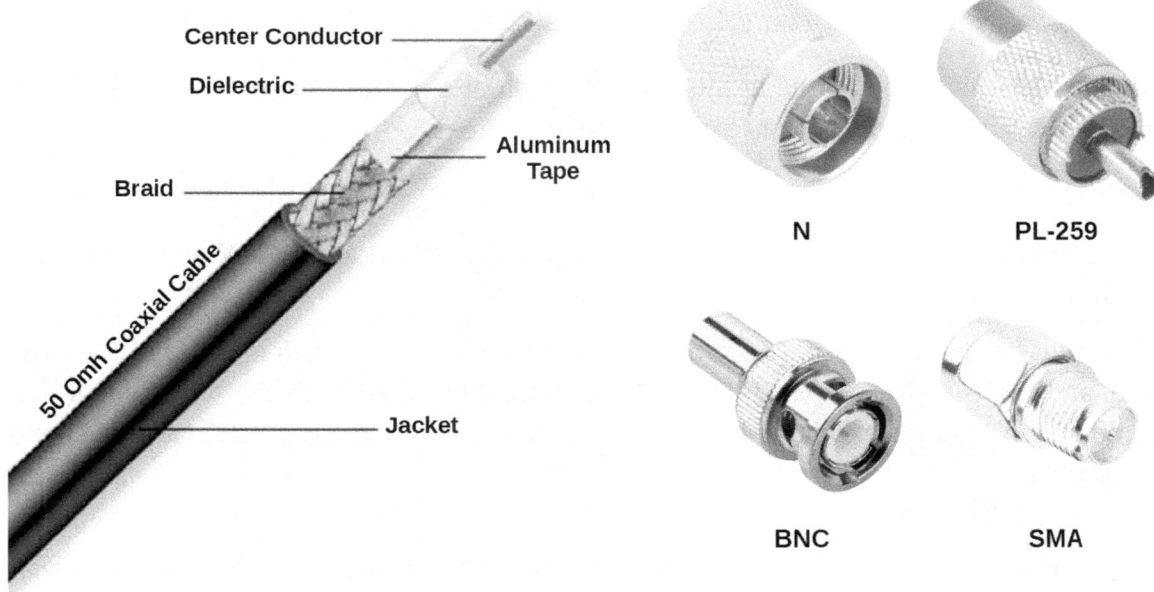

50 ohm Cables and Connectors	Max Transmit Power with <2.0 SWR			
Coaxial Cables	**<70 Watts**	**<200 Watts**	**<400 Watts**	**1500 Watts**
RG213/U	✓	✓	✓	✓
RG8X (More efficient than RG-58. Also called Mini 8/U)	✓	✓	✓	
RG58	✓	✓		
RG-58/U (Small diameter RG58)	✓			
Coaxial Connectors				
N (A threaded, weatherproof, medium-size RF Connector used up to 5000 watts)	✓	✓	✓	✓
PL-259/SO-239 (Also called UHF)	✓	✓	✓	✓
BNC (Bayonet style connector. More efficient than PL-259)	✓	✓		
SMA	✓			

All of the transceivers discussed in this book transmit at a power level of 100 watts or less and use 50 ohm cables and connectors for antenna connections.

Cable Performance

In the table above, three coaxial cables are shown, RG213, RG58 and RG8X and these standards work well for the transceivers discussed in this book.

RG58 is lousy older cable that uses solid polyethylene insulation which is not as efficient as the newer and more expensive RG8X. RG8X is commonly used with the transceivers discussed in this book. For even higher transmit power and less interference, RG213/U is a solid Mil Spec coaxial cable that is used for long antenna feedlines where you need to reduce the signal noise floor for base station transceivers.

However, for components installed in vehicles, RG213/U lacks the flexibility of RG8X. Also, the signal loss with RG213/U and RG8X are the same, so due to its wide availability

RG8X is generally preferred by radio manufacturers and will likely be your mainstay coaxial cable.

Cable Connectors

There is not much of a price difference between PL-259 and BNC connectors, but there is a difference with transmit performance. It is why the low-power handheld SOTA class and QRP CW transceivers presented in this book for field operations come with BNC connectors instead of UHF connectors because BNC is more efficient.

BNC connectors for these transceivers, which typically offer only five to ten watts of transmit power, can be upgraded in the field with external amplifiers to add another 10 watts or so of transmit power. Therefore, the highest possible efficiency combination for low power transceivers will be BNC connectors with RG8X coaxial cable for <200 watt transceivers.

The second advantage of BNC connectors and one that makes them very popular with HAMs is that they are much easier to connect and disconnect from components than PL-259 connectors. And finally there is a third type of connector you'll encounter called SMA. These connectors are typically used with amateur handy-talkie (HT) transceivers.

There is a wide range of adapters for converting between connector standards such as SMA to PL-259 and you will want to stockpile these adapters. The reason is, that over time your scavenging and bartering efforts will yield a number of serviceable transceivers of many different types and kinds.

Nonetheless, a good two-fold strategy is to avoid the unnecessary use of adapters wherever possible as they will marginally reduce the transmit power through your antenna. Then go with what the manufacturer installs on the transceiver. If it has a PL-259 connector, such is commonly used with 4 watt CB radios and amateur mobile transceivers, then stick with PL-259. Also keep in mind that PL-259 (UHF) connectors can be used with up to 1,500 watts of transmit power.

Now that we've identified all of the components of an example inline lightning defense, there is one last item. When do you shut down the station and how?

Emergency QRT Antenna System

QRT is a ham radio CW Q code for "shut down the station," and is typically used with the scheduling of off-air time such as a single radio operator closing a station down for the night.

During a global tribulation, this Q code will take on an added meaning, because there will be times when it is necessary to shut down the station on an ad hoc basis – and darn fast at that!

With this in mind, you need to implement an emergency QRT shutdown strategy to float your antennas when circumstances necessitate closing down an operational base station on short notice.

As we've already identified the equipment needed to build an emergency QRT system for your antennas, let's organize them into a complete system approach.

The key to this QRT system is two linefeed mounting plate pairs for floating. One above ground serving the antennas and another underground in the bunker for the base station. You can also design multiple QRT systems which all connect to the same grounding field.

The standard practice for grounding systems is to bond all connections to a single point grounding field. However, complications may arise due to your construction methods, location, soil condition, etc. and your technical team may opt for two widely-spaced grounding fields. One field would be for the antennas above ground and the other for a common grounding field for the underground shelters. Either way, the two basic variants of any grounding system intended for use during the global tribulation are:

- **Glow to Glow:** One QRT antenna system can be devoted to use with a "No Glow" policy where you stop operating the transceiver before the first glow of sunrise and after the last glow of sunset. Here, scheduled use is augmented with an emergency QRP disconnect.

Glow to Glow

- **24/7:** Another QRT antenna system can be devoted to continuous use, such as the radios and transceivers you'll use for field operations and security operations. Here, the risk of daytime operations is a given risk, and emergency QRP disconnects are performed only when a clear and present danger presents itself.

Regardless of your eventual QRT antenna system design, each underground component must be connected to the same grounding field and each must use the same design. This way, when the connections are unplugged on both sides, the continuity path from the antenna, through the coaxial feedlines that eventually connect to your radios is completely severed in two places.

Grounding System Connectors

While low-power two-way radios work best with BNC and SMA connectors, you'll want to use PL-259 (UHF) connectors and adapters to build your emergency QRT antenna system for the following reasons:

- **Cable Choices:** PL-259 (UHF) connectors and adapters work with all of the cables mentioned above which gives you a range of solutions including the ability to transmit up to 1500 watts. For example, you could use a mix of RG8X and RG213/U with this design. RG8X for <200 watts and RG213/U for >1500 watts.

- **Wide Availability:** The PL-259 (UHF) standard is popular and well established. The consequence of this is a wide variety of readily available connectors and adapters, and they will be in much greater supply than other types.

- **Specialty Adapters:** Two specialty PL-259 adapters you want to use are bulkhead adapters which will give you a way to connect to both sides of a linefeed mounting plate and push on adapters that replace screw on connections with push on (quick) connectors.

A critical component in this design will be the linefeed mounting plates. Like grounding field entry plates, these need to be copper as well. Brass would be your second choice and then aluminum for when you cannot acquire copper or brass plates.

Quick Disconnect

In this design, four linefeed mounting plates are used in two quick disconnect pairs. One linefeed plate is connected to the grounding field and fixed in place. The other is not connected to the grounding field and is not fixed in place. Rather, it has a rope safety lanyard that can be used to quickly pull the two plates apart.

In addition to the safety lanyard, a mechanical device with a lever can be used or a DC powered reversible screw driver. Either way, simple safety lanyards made of rope need to be incorporated into your final design.

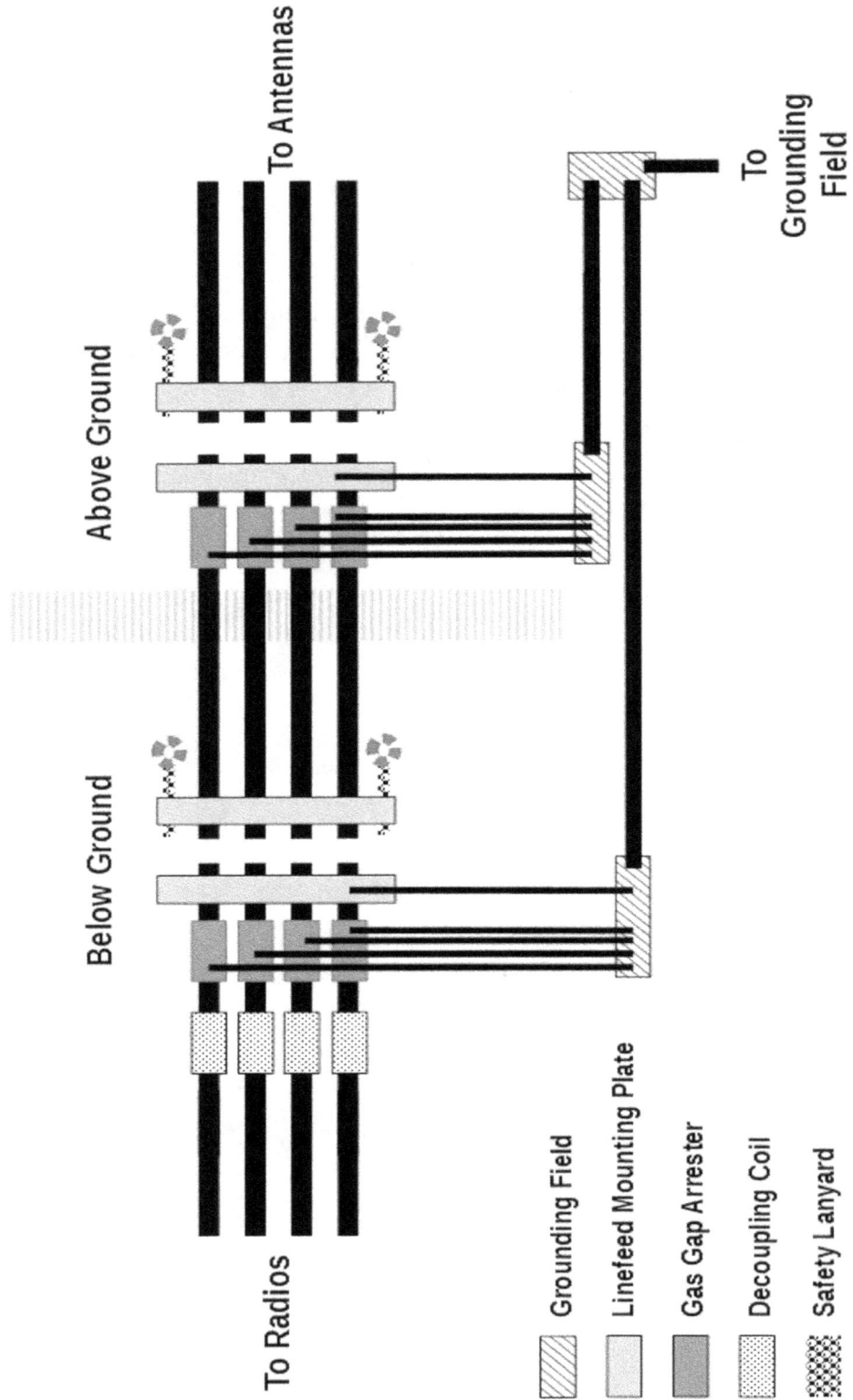

Emergency QRT
Antenna System

Glow to Glow or 24/7
Single Grounding Field

To Antennas

Above Ground

Below Ground

To Grounding Field

To Radios

Grounding Field
Linefeed Mounting Plate
Gas Gap Arrester
Decoupling Coil
Safety Lanyard

Once a linefeed mounting plate with safety lanyards has been disconnected, it should be moved as far away from the grounded linefeed mounting plate as possible. It is strongly advised that there is at least 6 feet of distance between the disconnected antenna linefeed connector and the feedline connector. Also, there should be provisions for safe storage so that no connectors make contact with the ground. By preventing damage from moisture and soil contamination, your feedlines will last longer and perform better over time.

For daily operations, you should designate two QRT shut down teams. One is for the base station, and the other is for above ground. They must be well practiced in rapidly disconnecting the components of your inline system and in the correct order. As a rule, you should begin to disconnect at the transceiver connection and then continue out to the antenna's feedline connection. For added team member safety, each should be issued a pair of lineman or electrical-insulating work gloves for added safety.

When all tasks have been completed, each team returns to a designated rally point and the leader verifies that all tasks have been successfully completed with a check list roll call. Assuming everything has been done according to procedure, the QRT is complete and the teams can be dismissed or reassigned.

In the next and last part of the book, Broadcasting and Training, the technical details for a local broadcasting system will be discussed, including additional grounding concerns.

Part 6 – Broadcasting and Training

20

Documentation and Training

In the previous parts of this book, we examined alternative communications and administrative mitigation as they apply to formulating an effective voltage spike mitigation strategy for EMP and lightning events.

These will be helpful in maintaining your operations but high standards for documentation and training also support your long-term operations, by prolonging the useful lifespan of your communications equipment during routine use.

Creating Tools

When HAMs buy a new transceiver, an instruction manual that manufacturers provide is a reference for operators with a basic proficiency and level of understanding of the relevant concepts. HAMs can then spend free time experimenting with their new acquisition.

This hobbyist HAM approach works fine because time is not of the essence. However, during a global tribulation time will always be of the essence. This means, community members will not have time to tinker and experiment. They need to use it properly from the get go.

Yet, this obvious fact seems to escape many who only want to buy inexpensive radios, box them up with batteries and store them somewhere until needed. That is, assuming they'll remember exactly where they stashed them or that they haven't moved, leaving their gear behind.

Therefore you may wonder. How will you know if someone felt it safe to buy something cheap without doing any documentation or training? You'll be likely to find them dead, lying face down in the dirt, grasping blood-smeared unopened boxes of communications gear.

Now you have to make a choice. Do you naively set yourself up to fail with a red herring assumption? Or, do you prepare your community for success using a standardized documentation and training strategy for each unique type of communications equipment, based on the following process steps:

- Know Your Users
- Mission Requirements
- Mission Profiles
- Reusable Content
- User Training
- Quality Control

This six-step process must always begin with each unique piece of communication gear and is repeated for all the other gear used by your community communications strategy.

To illustrate the concept of how good documentation and training makes a real difference, let's use a transceiver we've discussed at length in the previous chapters, the Kenwood TM-V71A Dual Band VHF/UHF mobile.

> **NOTE:** *This is not a universal example, because the features and functions available with different manufacturer brands and models will vary from transceiver to transceiver. Therefore the purpose of this example is to demonstrate documentation concepts based on one specific transceiver only.*

The first step in creating documentation and training tools for the TM-V71A is the same as any other transceiver. It is to identify who will use it and for which mission roles.

Know Your Users

The biggest mistake one can make when creating user documentation is the false assumption that the author is representative of the users. Never forget, this kind of "I am the world" thinking is reckless! This is why you need to get to know your users by talking to them about their needs and concerns, before you set pen to paper.

Therefore the first step is to broadly define your various user groups according to the three levels of your community communication strategy. In this case, the broadest possible user groups for the TM-V71A are:

- **Level 1 – Emergency System Radios:** Not applicable – no users.
- **Level 2 – Field Radios and Repeaters:** Possible use by teams operating in the field and as cross-band repeaters.
- **Level 3 – Command and Control:** Used in your command base station by senior personnel and HAMs.

As the TM-V71A is not intended for level 1 users, this means that your broadest possible user groups will come from levels 2 and 3. The primary difference between the two user groups are the mission roles performed at each level.

For example in level 2, the TM-V71A can be set up for use as a cross-band repeater by level 3 technicians for use by field teams operating beyond your community's outer perimeter. However, given that field teams may also need to configure a TM-V71A cross-band repeater in the field, the relevant documentation in this case is the same for both user groups.

Once you've refined your groups, the next step is to assess their level of proficiency. For field teams operating beyond your outer perimeter, the following user roles could be considered as examples:

- **Dedicated Field Radio Operator:** Ideally, each field team should have a dedicated radio operator who is fully trained and proficient in not only the TM-V71A, but with each of the other communications components used by the team as well.

- **Field Team Members:** While field team members are not expected to be as proficient as their dedicated radio operator, they need to know how to communicate with the community in the event of an emergency where the field radio operator dies or is unable to perform his or her duties.

- **Autonomous Field Operative:** In special circumstances, you could have a single, autonomous field operative operating in the field. In this case, the operative will need to be as proficient as a dedicated field radio operator and will likely carry just one or perhaps two transceivers.

With other circumstances, the skill level and experience of individual users is considered. However, for the process, you always proceed on the assumption that your users are minimally proficient, regardless of what they already know and can do.

Once you've created a detailed list of users for a specific component, the next step is to define the various mission requirements for that component.

Mission Requirements

Field operations can have multiple communication missions and each mission will have a unique set of requirements. It is also likely that there will be a high degree of overlap between different missions and their respective requirements.

Here, you move from the mission role of the user to the physical parameter of how each communication component is to be used in service to a field team's objectives. For our TM-V71A example, those could include:

- **Equipment Type:** Is this a handheld, mobile, backpack or SOTA-class transceiver? The TM-V71A is primarily an on-road transceiver, but can also be used for limited off-road and backpack configurations.

- **Bands and Frequencies:** Which bands will be used? The TM-V71A, in this case, will be used for the VHF and UHF amateur frequencies in the 2m and 70cm bands.

- **Operating Modes:** The TM-V71A transceiver can scan and receive frequencies transmitting in AM mode, but can only transmit in the FM phone mode.

- **Auxiliary Equipment:** The TM-V71A works well with a mounted whip or mag mount whip antenna. The field team may choose this transceiver over a MIL-STD model to hop ridgelines using amateur radio satellites and special Yagi antennas.

At this point, you will have identified the users and the communication components they will be assigned. In practice, there may be little variation from mission to mission in terms of user and equipment, but what will vary will be the mission profiles.

Mission Profiles

Amateur transceivers can be configured for different mission profiles, while simple transceivers such as a CB can only be configured for one mission profile.

The TM-V71A was designed for family use, but can be configured with five different sets of transceiver configurations, so that each family member can define and later recall and use their own configurations when working the transceiver.

In terms of survival missions, this family-oriented approach used by Kenwood can be easily adapted for specific mission profiles using different settings which are stored in the transceiver's Programmable Memory (PM).

Each configuration saved in the PM will store virtually all settings currently set on the transceiver when the configuration is saved.

PM Mission Profiles

Let's assume your technical teams need to configure a TM-V71A for five different PM mission profiles. While there are many settings that can be saved, let's just use a few to illustrate the point. These are:

- **Call Channel:** This is a handy feature because there is a single function key on the front panel of the TM-V71A labeled "CALL." The call frequency is like a national calling frequency or a community assigned tactical channel for initiating a contact. After the contact is made on the call channel, the conversation can be moved to another frequency to complete it.

- **Display Brightness:** The TM-V71A has eight different brightness settings from low to high.

- **Backlight Color:** Two display colors are available. Amber and green.

● **Key Beep:** The key beep can be set to on or off. The factory default is on, which means that when you press the PTT key on your microphone or a button on the front panel of the transceiver, there is a confirmation beep.

● **Key Volume:** This setting is used when the key beep is set to on. There are seven volume settings from 1 to 7.

● **Output Power:** Kenwood offer different international versions of TM-V71A. Those sold in the USA are designated as the K type and offer three different transmit power settings: Low (5 W) medium (10 W) and high (50 W).

Creating an optimal display configuration requires appropriate backlight color and brightness settings::

● **Daytime:** A bright amber backlight is a good initial setting for routine daylight operations.

● **Nighttime:** A dim green backlight is a good initial setting. This is because your eyes are more sensitive to green than amber, but do not use a bright green backlight, as it will have a negative impact on your night vision.

For years, the nighttime military standard for display background color was red. However, that changed to green with the advent of night vision equipment.

Now that we've identified five of many different types of configuration options available with the TM-V71A, let's create five different PM mission profile examples, each of which must be documented.

TM-V71A PM Profiles

We'll assume in this example that your field operations are conducted in two areas. One is to the East of the community, and another is to the West. Both areas have their own call channel and tactical frequencies. After experimenting with different configurations the following five are chosen.

| PM Profile | Call Channel | Display | | Key | | Output Power |
		Brightness	Backlight	Beep	Volume	
East Daytime	145.150	6	Amber	On	4	Medium
East Nighttime	145.150	3	Green	On	2	Medium
West Daytime	145.290	6	Amber	On	4	Medium
West Nighttime	145.290	3	Green	On	2	Medium
General	146.520	8	Amber	On	5	High

Amateur transceivers like the TM-V71A typically offer different ways to reset the transceiver. Some are non-destructive and used to clear current frequency settings for example, whereas another is a complete factory reset that eliminates all of the stored frequencies and PM profiles.

A factory reset should only be used by a technician when servicing the transceiver. However, if a field team sees that capture or death are imminent, the operator can use a factory reset to prevent valuable PM profiles and bandplans from falling into enemy hands.

This is how to determine what procedures the radio operator will perform for the mission and in which order, from beginning to end. This process is shepherded by your technical team for the purpose of publishing reusable content for your users and the missions they will conduct.

Reusable Content

After identifying the mission users, requirements and profiles, is it necessary to publish a field manual for each mission your field teams will conduct? No, because you will never have enough time to produce a series of user guides.

Rather, you need to set up a documentation system based on the traditional American Automobile Association

AAA TripTik Model

Before the internet age, traditional AAA TripTik paper maps were the road navigation aids of first choice. AAA members could obtain them at any local AAA office and these maps were simply awesome.

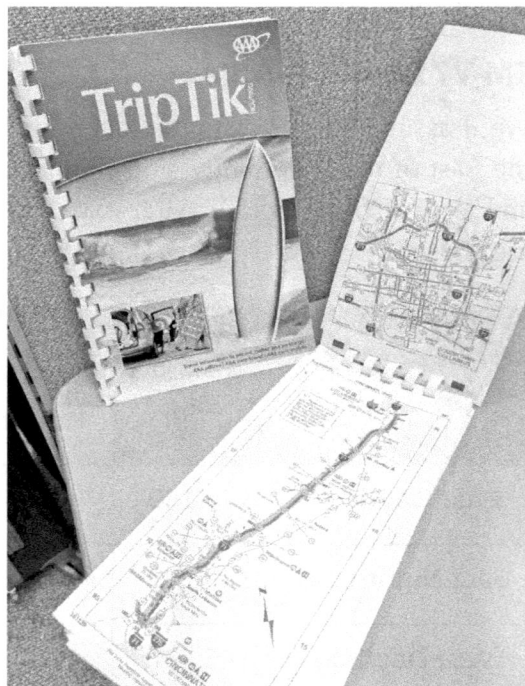

Today, you can no longer get a traditional Trip-Tik at a local AAA office, though they can be ordered from the central AAA office. Just keep in mind, you'll have to wait a few weeks for them to arrive so plan ahead. Meanwhile, as an AAA member you can immediately access online electronic navigation aids before and during your trip.

In the old days, AAA members explained their travel goals to a local office staffer and then watched that staffer pull handfuls of 4-by-9-inch paper maps from a long line of filing cabinets based on your trip preferences.

Each of these 4-by-9-inch map pages progressively spanned a short distance such as 50 to perhaps 150 miles and provided ample details about the route, nearby roads, interchanges, and where to find gas, food, and lodging.

Once the AAA staffer has gathered the necessary TripTik pages, they would then be sorted in travel order beginning with the departure city and ending with the final destination. Thus assembled, the staffer would then bind the stack with a spiral-bound spine for easy handling.

Today, spiral binding tools and supplies are readily available. However, during a global tribulation, something more durable and simpler will be necessary and here there is good news. There is a simple, inexpensive solution and airplane pilots have been using them for years. They're called pilot kneeboards.

A typical kneeboard will cost under $25 and will weigh about 10 ounces or less. It will likely be 6" x 9" in size and constructed of thick aluminum board with foam backing. It will also have spring loaded clips for holding multiple paper maps and elastic straps for attaching the kneeboard to the thigh. These kneeboards can be purchased in volume and be assembled by your own community members.

Like the traditional AAA TripTik, you can create multiple, reusable paper pages. Each must be a single procedure checklist card written specifically for each brand and model of transceiver included in your community communications strategy.

Procedure Pages

To avoid confusion, you want to use common procedure pages and descriptions. For example, the three procedure checklist descriptions below can be universally applied to all brands and models:

- **Check Out:** When any radio operator or field operative checks out a transceiver, they need to use a checklist that walks them through the hardware connections and programmed functions so the transceiver is confirmed as being deployment ready.

- **Check In:** When turning in a transceiver following a field operation, a checklist walks the operator through the proper disassembly and storage of the unit.

- **Radio Reset:** In some cases, such as lightning strikes or static electricity a transceiver's programmable configurations may become corrupted or not work as well as previously. However, resets should be used sparingly and only after other troubleshooting efforts fail to resolve the problem.

  Resetting a simple transceiver such as a CB could be as simple as turning it off and on. However, amateur transceivers are more complex and typically offer multiple reset options.

When writing a procedure page, there are seven things you can do when organizing your mission profile procedures.

- **Card Stock:** You need something durable for the field and copier paper is not the first best choice. Rather, you need card stock, such as the kind used by trade paperback book covers, which is also called cover stock. This type of pasteboard is thicker and more durable than normal writing or printing paper, but thinner and more flexible than heavier types of paperboard.

- **One Card Per Procedure:** Each procedure must fit on one card and include both the front and the back. When using a 6" x 9" kneeboard, a readily-available card stock will be 5" x 8" index cards.

- **Legibility is Essential:** Whether you are printing procedure cards with a device or by hand, legibility is essential. Using a small font to save space will only make a procedure card difficult to read in adverse field conditions. Also, highly legible, printed lettering must be used for hand-written cards.

- **Standard Formatting:** Develop and use consistent headings, terminology, organization, and verbiage and always remember the first rule of documentation. "Be consistent."

- **Define Acronyms and Terms:** It doesn't matter if you feel it is redundant; always define each acronym and term used with a procedure. This is necessary because undefined terms or acronyms can create confusion.

  Remember, people only remember about ten percent of what they read. Therefore, the assumption that the user will remember each acronym or term defined elsewhere is wrong. It will only create or worsen a needless crisis.

- **Simple Language:** When creating procedure cards you will likely need to refer to the instruction manual the manufacturer shipped with your transceiver. These explanations are principally written for a technical audience, where there are assumptions about what the reader understands. This is why newbies are often confused by factory instruction manuals, or simply give up in favor of a clumsy trial-and-error process.

- **Keep It Simple:** A common aspect of amateur transceivers is that there is more than one way to configure an option and so instruction manuals will offer a range of options. A tactic that suits experienced HAM operators – not newbies, especially when they are operating in the field beyond your outer perimeter.

For the reasons given above, always use the most simple and direct way to execute any procedure when authoring documentation and remember that you're performing a work in progress task, one that is improved over time through user training and quality control.

User Training

Earlier, we discussed the need to have a systems analyst / senior technical writer on your technical team. If a trained writer is not available, a high school teacher with solid experience in developing teaching plans can fill this role.

Nonetheless, whomever you charge with this responsibility, instruct them to use the following five-step approach when creating documentation and training tools. Also, make sure they report back to your technical team on the status of their efforts based on this five-step approach.

- **Learn the Component:** Before you can document any procedure with a given component, the author must be fully acquainted with the component's features and operating modes. This will be a combination of HAM guidance, self-instruction using a manufacturer instruction manual and hands-on experimentation.

- **Interview Potential Users:** Authoring a book or a procedure is much like giving a speech. For those who are nervous about getting in front of a group to make a presentation, public speaking trainers suggest finding a friendly face in the middle of the audience; then, always speak to that face. When writing mission procedures, the same advice applies. Find one user to serve as your initial user, such as team leader or field radio operator to consult with before drafting the first versions of your procedures.

- **Draft and Review Procedures:** When establishing a working relationship with an initial user, ask them about what they believe they need, and what they feel will work best and then draft the first versions of your procedures. After that, review the drafts with the initial user plus other community members in various roles and make whatever revisions are necessary. Then revise and review your documented procedures on an ongoing basis.

- **Test the Procedures:** Training newbies will be the best way to test your procedures for clarity and ease of use. Always be listening for what technical writers call dumb user questions, such as "you mean I have to plug the computer into a wall outlet?" As funny as they may sound, dumb user questions can flag critical gaps or ineffective explanations in your documentation, so treat them with respect and gratitude.

Of course, you can expect members to voice their concerns or to make suggestions about your documentation and training. Do not take this input personally nor should you allow anyone on your technical team to do so. Rather, you need to use a quality control process to continually update and improve your documentation and training tools.

Quality Control

Assuring quality control is a straightforward process. At every opportunity, your technical writer learns what connects with operators when conducting training and by debriefing field teams immediately after they return to base and check in their communications gear.

The questions need to be simple, such as:

- Overall, how well did the documentation work for you?
- Are you confident in your communications gear and you ability to use it?
- Did you have any difficulty finding the information you needed?
- Is there anything we can do to improve the documentation?

⦿ Are there any suggestions or ideas we should explore?

How often do you ask these questions? Every chance you get and you never stop asking because the environment during a global tribulation will be in a continual state of change.

Also, always keep a log book on everything you do. Not just to avoid repeating mistakes, but when circumstances change, you have a resource of information from previous efforts.

Granted, what this chapter is asking you to do is something that most feel is boring, but what is the alternative? Losing valuable equipment or lives because your operators were desperate for a solution and couldn't find one? This is what happens when something as ordinary as documentation is left to complex reference manuals and desperate attempts to find a solution.

We're not talking about rocket science here or making contacts with others hundreds of miles away. Rather, we're talking about simple things that can go wrong from the beginning.

To illustrate this point, let's look at the example of transmitting without a load damaging a transceiver due to a lack of a committed documentation and training strategy.

Eliminating Confusion

The term "transmitting without a load" refers to what happens when a transceiver is powered and connected to a microphone and a power source, but the antenna connector is exposed without a connected antenna or dummy load.

Do manufacturers anticipate these kinds of needless accidents? Sometimes they do with newer models and a safety circuit is built in to automatically prevent this kind of damage. However, how can you know a particular transceiver has this built-in feature since this information is seldom mentioned in an instruction manual?

Then on the other hand, a seasoned HAM operator will tell you that you can safely power up a transceiver without connecting it to a matched antenna or dummy load without damaging it. This is absolutely true, because the only way damage that can occur is when the push-to-talk (PTT) switch on the microphone has been depressed or the radio has been keyed after the transceiver is turned on.

For HAMs this makes perfect sense, but for a newbies, explanations like these are full of hobbyist technicalities, exceptions, and work-arounds that can easily create confusion that leads to a needlessly damaged transceiver.

What is the best way to prevent needless damage to expensive and valuable communication gear? Good old KISS – keep it simple stupid. So instead of a confusing hobbyist explanation, here is the right approach to this problem for newbies.

Create a check out procedure card that describes how to connect the correct power source, microphone and antenna for each transceiver with a huge warning in large red letters.

"The antenna supplied with this transceiver must be connected before switching it on."

Yes, for a HAM, this may be simplistic overkill, but for a newbie, it is simple, straightforward and safe. Therefore, when creating documentation, never forget, it is far better to be safe than sorry.

Effective documentation and training is essential to the survival of the community, and it must be the responsibility of every community member. Also, there must be universal proficiency of a basic set of communication survival skills.

21

Universal Proficiency

Suggestion: Universal proficiency in general preparedness skills will be needed. Marksmanship, tying knots, purifying water and so forth are necessary community survival skills. Likewise, your community communications strategy will also require that each member of your community is proficient in basic, universal communication-related skills.

It makes no difference whether a community member is a HAM or a trained field radio operator. Or, for that matter if that community member will only use a two-way radio under emergency circumstances. For this reason, it is imperative that every member of the community is proficient in basic, universal communication-related skills.

Land Navigation

In the days before the Global Positioning System (GPS) basic land navigation tools were used. Land navigation tools are simple devices such as topographic maps showing terrain features and lensatic compasses (also known as military compasses) with magnifying lens for reading the scale.

During a global tribulation, a back to basics strategy will be necessary for field teams operating beyond your outer perimeter. This may prompt some to ask, "Why do I need this? I already have a GPS positioning app in my smartphone? Even if the grid goes down I still have the app."

The answer to that question, "you have nothing."

This is because smartphone GPS positioning and navigation apps require an active cellular system to work. If you still want to use the GPS system, assuming it is still operational during a global tribulation, you'll need to use dedicated handheld GPS receivers. Unlike a simple, reliable lensatic compass, this means that you will be dependent upon batteries to power your device, which contains electronics that are also susceptible to EMP events.

What will be more prevalent during the trying times of a global tribulation will be the use of simple hand-drawn maps with reference landmarks and community naming conventions.

For example, a double peak to the West of the community will appear on a topographical map, but you'll want to create your own names for recognizable landmarks and terrain. Ideally, you'll want to define locations or terrain features that are recognizable from 360° around them.

In this case, the double peak could be called Lucy's Saddle for example. This way, instead of transmitting map coordinates for this formation in the clear which may put field teams at risk, you use Lucy's Saddle or whatever else you've named it within your community.

The same policy will hold true for other locations used by your community members for which there will be different needs such as rally points for field teams and member emergency areas.

Rally Points for Field Teams

For those with field combat experience, the use of rally points is a given and when assigning rally points, it is important to designate their type. The military has several types, but for survival communities these may not an ideal fit. Rather, designating survival types based on your location and situation will likely be more suitable such as:

- **Initial:** An egress point near your outer perimeter.
- **En Route:** Various rally points along a given route based on terrain, visibility and so forth.
- **Objective:** When operating beyond the outer perimeter, a final rally point in the safest location near the field team's objective.
- **Reentry:** A rally point outside of the outer perimeter where the team can assemble before returning to the community.
- **Near and Far Side:** Rally points to either side of a dangerous area.

If your community must abandon its location or is scattered, these rally points will be helpful. All members should know the locations of the emergency areas.

Member Emergency Areas

Physically fit members who can carry an 80 lb pack and shoot the eye out of a squirrel at 50 yards will be invaluable and there's no doubt about that. However, so will a grandmother with arthritic hands. She may not be able to comfortably pick weeds, but she'll know every medicinal plant in the area and how to prepare it and of course, there are the children. They will be the crown jewels because in them will be hope for the future and a sense of purpose.

For these reasons, it is essential that every member of the community, especially the children, know where the member emergency areas are located and how to access them. These can vary with circumstances, so here are a few member emergency area types that illustrate the point:

- **Secure:** Additional shelters inside the community's inner perimeter. A good example would be an above ground entrance to a small underground shelter for protection from meteor showers, solar storms, hail, tornadoes and so forth.

- **Control:** Between the community's inner and outer perimeters. Typically, trenches, observation posts, and camouflaged sniper positions.

- **Operating:** Areas where your field teams typically operate within the line of sight of your outer perimeter.

- **Spider Holes:** Small secure temporary hideouts for protection from natural threats and concealment from attackers.

- **Messaging Spots:** Any area where community members can post a hand-written emergency message for other community members.

- **Evacuation:** Emergency assembly areas beyond line of sight of your outer perimeter. Typically used if the community is overrun by attackers or becomes untenable due to natural causes.

In an emergency situation, expect members to use two-way radios in phone mode at first and whether it is AM, FM, or SSB makes no difference. The message intelligence is what matters and here is where you need a way for a hysterical or wounded member with limited two-way radio proficiency to be able to clearly announce their situations, location and needs over the air.

With these situations in mind, everyone in the community (not just field teams and HAMs) should know the NATO Phonetic Alphabet and the 24 hour clock by heart.

Phonetic Alphabet and 24 Hour Clock

An important survival skill is the ability to phonetically spell words using the NATO Phonetic Alphabet. Used with military and aviation voice traffic, HAMs also rely upon this standard International Phonetic Alphabet.

Letter	Phonetic	Letter	Phonetic
A	Alfa	N	November
B	Bravo	O	Oscar
C	Charlie	P	Papa
D	Delta	Q	Quebec
E	Echo	R	Romeo
F	Foxtrot	S	Sierra
G	Golf	T	Tango
H	Hotel	U	Uniform
I	India	V	Victor
J	Juliet	W	Whiskey
K	Kilo	X	X-ray
L	Lima	Y	Yankee
M	Mike	Z	Zulu

What makes the NATO phonetic alphabet so useful for survival communications is that it can help prevent spelling mistakes or miscommunication with local radio traffic. For international traffic on the HF spectrum region, it will also help you to communicate with people from different countries who speak English as a second language.

In addition to the NATO Phonetic Alphabet, each member of your community should also understand and use the 24 hour clock, also known as military time.

Military	Civilian	Military	Civilian
0001	12:01 am	1300	1:00 pm
0100	1:00 am	1400	2:00 pm
0200	2:00 am	1500	3:00 pm
0300	3:00 am	1600	4:00 pm
0400	4:00 am	1700	5:00 pm
0500	5:00 am	1800	6:00 pm
0600	6:00 am	1900	7:00 pm
0700	7:00 am	2000	8:00 pm
0800	8:00 am	2100	9:00 pm
0900	9:00 am	2200	10:00 pm
1000	10:00 am	2300	11:00 pm
1100	11:00 am	2400	Midnight
1200	Noon		

Interestingly, the 24 hour clock is the most commonly used interval notation in the world and for tribulation survivors who will likely spend days at a time underground, it will eliminate the confusion caused by the civilian AM and PM time notation system.

The universal proficiency recommendations discussed above will enable the rapid exchange of signal intelligence (the content of your message) without needless confusion.

However, even when community members have this proficiency, being clearly understood while under pressure is best achieved with the proper microphone technique.

Microphone Technique

Previously, we discussed how CB radio operators would use aftermarket powered noise canceling CB microphones, such as the DM-452 Power Echo Microphone, in order to be heard like a "big audio" station.

The problem with these types of microphones is that they can cause distortion and when used incorrectly they create something called splatter or splash over. Splatter occurs when an overmodulated signal bleeds over into other channels and disrupts other traffic. This happens when a powered microphone is configured with a power level that is too high which is why the use of powered CB microphones is not advisable during a global tribulation.

Microphone technique is a skill that applies to every type of two-way communication and the three principal types of microphones your survival community members will use most often will be:

- **Base Station Microphone:** Also known as desktop microphones, base station microphones are designed for use with a base station transceiver. They are not intended for mobile use in the field, but are excellent for mobile and portable transceivers used in a base station.

- **Built-in Microphone:** Amateur handy-talkie and consumer walkie-talkie handheld transceivers have built-in microphones on the front face of the transceiver.

- **Hand Microphone:** Amateur and CB/SSB CB mobile and portable transceivers typically come from the factory with a hand microphone that may or may not have a built in speaker. Aftermarket versions can also come with built-in speakers.

There are other types as well, such as earpiece microphones which security personnel often wear. With these other types, microphone technique is automatic due to the way they are worn and used. However, for the three types mentioned above, there are two primary aspects of good microphone technique: distance and angle.

Microphone Distance

Distance is a universal concern, because if your mouth is too far from the microphone, your audio transmit level will drop significantly. This in turn will cause your spoken message to be difficult to understand, especially when there is noise or static at either end.

Likewise, the dynamic range of your voice will also be impaired and if you have a weak radio voice, this could make you difficult to understand over the air.

A cheap and easy fix for all this is to compensate for distance issues is to use a powered microphone. Wrong! Again, this is misleading logic which also has some serious downsides. A powered microphone will not only make your voice louder, it will likewise make all of the distracting sounds around you louder as well.

Another wrong-minded easy fix is to hold the microphone extremely close to your mouth. This will only result in wet mouth sounds and/or noise from your face, beard or lips brushing against the microphone and you'll sound like a bad imitation of the Star Wars character, Darth Vader.

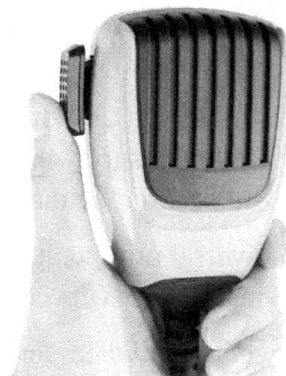

So what is the proper distance? Regardless of which type of microphone you're using, your mouth needs to be 1" to 2" away from your mouth. The other aspect of good microphone technique is angle.

Microphone Angle

Each of the three primary types of microphones described above will require a different angle relative to your mouth.

- **Base Station Microphone:** Typically used by radio dispatchers and HAMs, these desktop microphones are typically positioned on base station desk in front of the transceiver and are pointed at the chin. Most are high quality cardioid designs with heart-shaped pickup patterns and can sit directly in front of you or at a slight angle to either side.

- **Built-in Microphone:** With a handheld transceiver you never talk directly into the microphone. Rather, you hold the transceiver at a 45° angle to your mouth and speak across the microphone. Not straight into it.

- **Hand Microphone:** Whether a hand microphone has a built-in speaker or not, you always hold it at a 90° angle to your mouth and speak across the microphone. Not straight into it.

How will you know if your microphone technique is poor?

With built-in and hand microphones, when you hold them close to your mouth and speak directly into them, your transmission will be peppered with what is called "popping."

Popping is caused by the impact of fast-moving air on the microphone pickup. It typically occurs when the operator is speaking explosive words, especially those starting with the consonants P and B. An easy way to test your microphone technique for popping is to use a popular tongue twister.

Peter Piper Microphone Test
Peter Piper picked a peck of pickled peppers.
A peck of pickled peppers Peter Piper picked.
If Peter Piper picked a peck of pickled peppers,
Where's the peck of pickled peppers Peter Piper picked?

If you can recite Peter Piper with a built-in or hand microphone with a normal voice and there is no popping in your transmission, this means the microphone is at the correct distance from your mouth and that you're speaking across it at the proper angle. For base stations, the microphone angle is not as crucial as it is with the other types.

Nonetheless, before assigning transceivers to your survival community members, instruct your technical team to test each brand for audio quality with different mic angles and distances.

When your community members remember and practice good microphone technique, you will enjoy more precise communications without unnecessary misunderstandings.

Another thing everyone in your community should have is a basic proficiency with the communication protocols two-way radio operators use.

Learn the Protocols

Newbies are often stymied by two-way communication protocols, such as how to announce yourself. While there are useful lists of terms for these protocols, as the saying goes, "some things are better caught than taught." This is why the best way to learn the protocols is to listen to contacts between other HAMs or operators.

For newbies who need to get on the air before mastering the protocols, here are a few basics:

- **Who You're Calling:** If you are calling a specific HAM, you'll use their call sign and for a CB operator their handle. To start a general conversation, use the term CQ which means "attention all stations" or just say attention all stations.

- **Who You Are:** Identify yourself with your call sign or handle and your general location such as the city and state. Never give your precise location to strangers.

- **Purpose of the Contact:** The purpose can be as mundane as a communications check to test your equipment, or a request for emergency services. Whatever it is, be concise.

- **Copy That:** When you receive a message from a contact, you can acknowledge receipt of that message by saying "copy that." This helps the other operator know that you're ready for the next part of their transmission, or to begin your response.

- **Over:** When you say "over" your contact will understand that you've finished your statement or response and now you are waiting for the contact to reply.

- **Out:** When you say "out" your contact will understand that you've finished talking and you're now leaving the conversation.

One important note on protocols: A favorite line in Hollywood movies is "over and out." Use this line on the air and you'll sound as dumb as a box of rocks.

These two terms have different meanings and should not be used together. You either say "over" or "out" but you never say "over and out." Another helpful tip is to speak with a purpose.

Speak with a Purpose

The protocols used to communicate over the air with two-way radios present a common challenge for newbies. Consequently, it is difficult for them to make contacts for idle chats. This is because amateur radio is not about talking to people as we do with smartphones and VoIP applications like Skype.

In other words, gossiping over the fence may be fun with a smartphone where you have a private channel, but not in the world of two-way communications. This is because you will dominate a shared public frequency with gossip that is unprofessional and disruptive.

On the other hand, if you are mindful of what is known as "HAM etiquette" you will be treated with courtesy and patience even if you are not proficient with protocols.

HAM Etiquette

The difference between smartphone gossip and two-way radio communication is purpose and HAMs prefer to communicate with a clearly defined technical purpose in mind. These include equipment testing, contest competition and emergency communications.

A few basic and helpful HAM etiquette protocol techniques for newbies include:

- **Be Authentic:** Always be yourself and speak with a natural voice. Remember, your voice will become recognizable over time when transmitting in phone mode. Likewise, with Morse code your "fist" (personal keying style) will also be recognizable. Either way, if you try to impersonate someone else or you avoid an organic tone in the conversation, your machinations will be noticed by others and you will be branded as a predator or an idiot. A hapless fate that is hard to come back from.

- **Be Professional:** A big distinction between the trash talkers often heard on the Citizens Band and the professionalism of licensed HAMs is effectiveness. For this reason, professional conduct is the best way to communicate your messages in the least amount of time and tent flap mouth nonsense. This is because courtesy, competence and respect go a long way with those who value it. As for those who do not, they will not value other things as well, such as the lives of your survival community members.

- **Drop the Urban Lexicon:** Foul language, filler words like "um," "like," "you know" and "err" and net-centric abbreviations like "OMG" and "WTF" have no place in two-way radio communications. They only make you sound incredibly crass, stupid, and naive. Worse yet, it sends a clear message to predators listening in – as in "I'm easy pickings because I'm stupid and confused."

- **Think Before You Speak:** There is a popular culture saying – "engage brain before putting mouth in gear." In terms of two-way radio communications, this translates to "have your thoughts and sentences clearly in mind, before you press the PTT microphone key and begin transmitting." Remember, pondering what to say as you transmit may create unnatural pauses and disjointed messages that make you difficult to understand.

When you employ good HAM etiquette techniques, you'll find HAMs are very supportive and patient with newbies who are learning the protocols of whichever service they are using.

Always remember, learn to think before you speak, as it could make the difference between life and death in an emergency situation. With this in mind, there are two other things that will help newbies be better understood, a clear enunciation and pacing.

Enunciation and Pacing

The goal of proper enunciation and pacing is to express your ideas and statements with a consistent focus and volume.

- **Enunciate Clearly:** Pronounce every word clearly and never let your voice drop off at the end of a sentence or thought.

- **Pace the Message:** Speak your thoughts and complete sentences in a relaxed, natural and paced manner to help the listener follow what you are transmitting.

The important thing is that your message is consistent and that you never trail off at the end with an unfinished thought or sentence. A good way for newbies to visualize proper enunciation and pacing is with a technique I call "Every word is a pearl."

To visualize this technique, imagine the each word, code or acronym in your message is a beautiful natural pearl and that your complete thought or sentence is a collection of pearls (words) stranded together to form a complete necklace, where each one is uniquely identifiable.

At this point, you now know how to speak with a purpose with HAM etiquette, clear enunciation, and consistent pacing. People will have an easier time understanding you when you do these simple things. However, if you want to make it as easy as possible for others to listen to you over a two-way radio, there is another technique you need to train yourself to "talk in the tube" as I call it.

Talk in the Tube

Broadcasters and professional entertainers have what some call "stage voice" or "radio voice" because they control their breathing and use their vocal cords quickly and with well-controlled tension. These are learned skills that take time to master.

For two-way radio communications, a stage voice or radio voice is not necessary, because this form of communication is not about entertainment or infotainment. Rather, it is about sharing information during a global tribulation. There will be no time for warm up exercises because you need to sing a song, tell a joke, or deliver the news.

Rather, there is a simple way for every survival community member to significantly improve their speaking voice with a self-help technique I call "talk in the tube." This technique addresses the following four basic concerns:

- **Bandwidth:** When we speak to each other face-to-face we hear each other with full fidelity, or in other words, full bandwidth. However, when we speak over two-way radios, we are dealing with much narrower bandwidths.

- **Attack:** When communicating a thought or sentence over the air, newbies tend to begin in a strong voice with loud volume. This is called the attack and a good attack keeps you within the bandwidth of the frequency you're working. If your volume exceeds the bandwidth limits, clipping becomes a problem.

- **Clipping:** When newbies begin speaking with a strong attack, there is a tendency for clipping. Clipping occurs when the attack is so strong, it exceeds the bandwidth. Consequently, the very beginning of a thought or sentence may become unintelligible.

- **Diminish:** As we finish communicating our thought or sentence over the air, it is natural for the strength and volume of our voice to reduce. This is called the diminish, and the more reduced it becomes; the more difficult it will be for others to decipher what is said.

If you are thinking that a human voice is just as clear over the air as it is in person, you're following a misleading assumption, as we communicate face-to-face in multiple ways.

First, we hear each other with full fidelity, so when the speaker is difficult to understand, we can fill in the gaps with non-verbal body language and context. Ergo, this is how we can decipher a weak speaker on a face-to-face basis.

However, when speaking over the air we always work in more constrained circumstances and here is where one simple technique can resolve all of the problems. It is called controlled breathing.

Controlled Breathing

Most people these days are what are known as "chest breathers." Their breathing is shallow because they're using their upper chest. Hence, they must shrug their shoulders and neck muscles to inhale. The result is very shallow breaths which degrade voice quality.

Were we all born as chest breathers? No. Observe a sleeping infant and you'll see their belly rises and falls. This is because each of us was born a belly breather and we just got away from that as we aged. Conversely, dogs and cats are born belly breathers and stay that way all their lives. Therefore, we human adults are short-winded chest breathers. That is, until we learn controlled breathing.

With controlled breathing, we're breathing from the belly or the diaphragm. Instead of filling their upper chests with shallow breaths, belly breathing techniques take us back to our infancy by teaching us to breathe deeply by drawing deep breaths of air into our entire chest.

This is a trade secret for all sorts of broadcasters, entertainers, and news announcers, and it is essential for their ability to speak with wonderful golden tones and to be easily understood.

What does it take for them to learn this? A lot of coaching, daily exercises, and practice. Obviously few of us if any will have time for this kind of training during a global tribulation.

However, there is a simpler way to become a proficient belly breather and thereby become easier to understand over the air. The easiest way to learn belly breathing for optimal over-the-air communications is to listen to your recorded voice.

Self-Paced Instruction

If someone in your community is an experienced voice coach, or a Hatha Yoga instructor, they can teach controlled breathing. Otherwise, there are ample resources for controlled breathing methods and once you've found one you like, you can use to train yourself with self-paced instruction.

Here is the good news with controlled breathing for two-way radio communication. There is an easier and faster way than personal coaching. You train yourself.

What is required is a computer or laptop, a microphone and an audio waveform editor, such as the freely-available Audacity Audio Editor (www.audacityteam.org). There are other free and paid audio editors as well.

How you train yourself is simple. Configure your audio editor with bandwidth settings such as those we see with defaults such as "voice memo" quality with the following settings:

- Sample Rate: 16000

- Resolution: 16 bits per sample

- Channel Format: Mono

Before recording your voice, make sure you are using proper microphone technique, then begin reading a sample paragraph with a mix of complex and simple sentences. When you are finished, save the recording and it will be displayed in a waveform with the audio editor.

When you record with a good radio voice, here is the waveform you want to see with your audio editing software:

Good Radio Voice

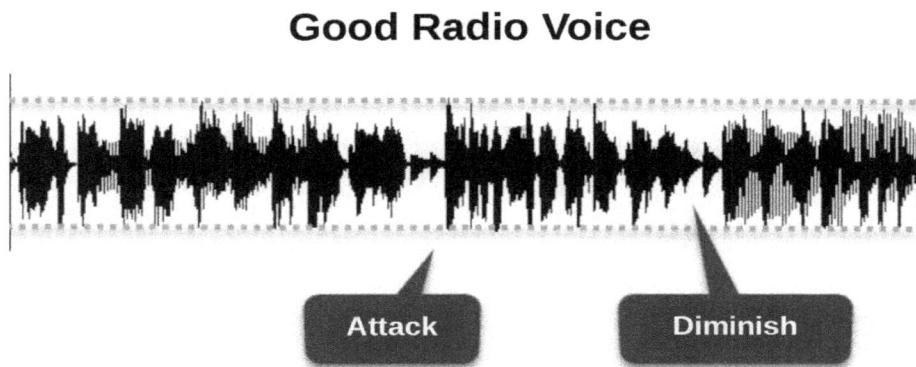

The illustration above shows proficient controlled breathing. The signal intelligence (the message) fills the bandwidth without clipping or a weak diminish.

When you see a tube-shaped waveform like this, you've mastered your controlled breathing, or as I like to say, "You're in the tube."

Conversely, what will an audio waveform look like for newbie chest breathers?

Poor Radio Voice

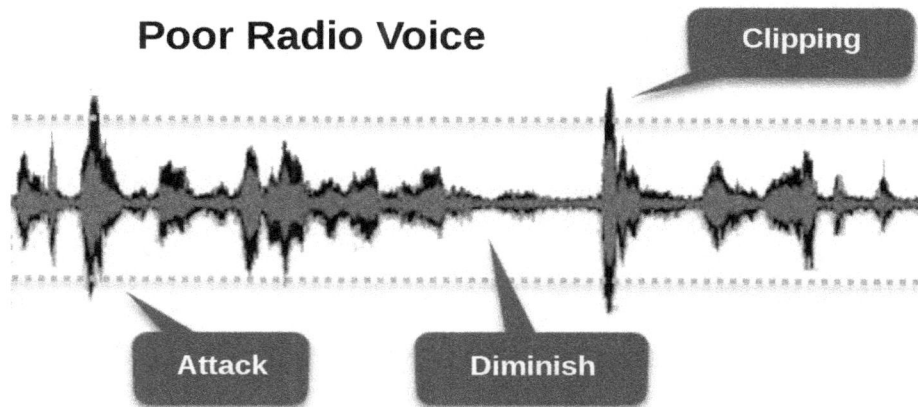

Below we see the waveform for a typical newbie chest breather.

First, look at the difference between the attack and the diminish with this poor radio voice illustration. What we see here is an over-modulated attack that results in clipping. This in turn causes the beginning of a thought or sentence to become difficult to understand or completely unintelligible.

Then we see that the waveform pattern following the attack plunges to a very weak volume as the thought or sentence diminishes towards the end. The result of this loss of volume in the diminish is that whoever is listening to the transmission will have to decipher it as it trails off. This is a great way to create misinformation and life or death errors that may come back to haunt you with a vengeance.

The point here is that once you find and master a simple controlled breathing technique, you will no longer shrug your shoulders and neck muscles to inhale because you're a chest breather. This means that your on air voice will be consistently steady and strong.

22

Knowledge Exchange

During the early stages of the tribulation, many survivors will use two-way radios to communicate with others for any number of reasons. It will be a chaotic phase and there will be much suffering before things settle down to a more stable existence.

Survivors during this time will adapt to these new circumstances, the end of life as we know it, and recognize that this misfortune is one of global consequence. Therefore, there will be no return to the earlier times of relative comfort and ease.

Yet, in these very circumstances, hope can be reborn. As the old saying goes, "Every dark cloud has a silver lining." The same is true for a global tribulation. What is the silver lining? It is the absence of control by the elites and suppression by the other powerful special interests.

Many have heard the stories about the people who have invented wonderful things. However, when they attempted to make them public, they ran up against Energy or Pharma special interests, and then what happened?

Special interests spot these inventors early, and when they submit working prototypes with their patent filing, their lives are targeted for suppression and suffering. Powerful special interests then use their money and political clout to eliminate any threats to their status quo. This is when the government goons invade the homes of honest inventors to take their work and wreck their lives.

In the end what do the special interests gain? Business and profits as usual.

But what do the rest of us lose?

We lose those groundbreaking inventions that could power our shelters and heal us from the myriad of new diseases that will be unleashed by a global tribulation. In other words, we lose the next generation of genius inventors and here is a real life example that makes the case.

Edison vs. Tesla

As mentioned previously, it was Nikola Tesla who invented radio according to the US Supreme Court – not Marconi, even though he is regarded as the father of commercial radio.

However, the real evidence of special interest suppression is found in Tesla's nemesis and former boss, Thomas Edison. The two feuded over which electrical system would eventually power the world. Tesla's alternating-current (AC) system won and Edison's rival direct-current (DC) electric power lost. Then as history shows, the feud became bitter.

Edison was an iconic inventor that many believe perfected the process of invention with a large facility and a sizable staff to handle the more mundane tasks. The result is that we now have light bulbs, movies and more. However, Edison's inventions all had one feature in common, a built-in market demand. If he hadn't invented those things, others would have eventually.

In contrast, Tesla was a futurist that marched to the beat of a different drummer as the saying goes. As a result, his inventions often became disruptive technologies. When Tesla invented AC, it disrupted the use of Edison's DC generators. This resulted in Edison doing what any wealthy special interest would do. Edison chose to put his own interests above those of the country, if not the whole world.

Fueled by vengeance, Edison unleashed a constant political and economic suppression to destroy Tesla's legacy and he came close to achieving it. Yet who's the one revered today Edison or Tesla?

Edison's abuse of Tesla, in an odd twist of fate, has resulted in a reversal of legacy. Fewer people now think of Edison as the "Wizard of Menlo Park" as he was once named, but as a fraud and a cheat. Conversely, Tesla's legacy is enjoying a huge comeback.

Why is Tesla so loved after decades of obscurity? It is because Tesla devoted himself to helping humanity and he spent years working on ways to wirelessly transmit voices, images and moving pictures. Tesla's inventions have become the backbone of our modern power and communication systems. Tesla was in essence the father of the information age.

Was Tesla a one shot deal? No. These wonderful, inventive geniuses are being born all the time and that fact will not change during a global tribulation. What will change is that we'll see a whole new generation of geniuses who no longer have to worry about powerful special interests and their government goons kicking in their doors for no lawful reasons.

Herein is the silver lining in this dark cloud of tribulation.

The brilliant inventions of talented men and women who want to share their discoveries and inventions with others will be free of suppression. When this happens, survivors will begin to realize how devastating the suppression by special interests was.

However, this newfound freedom from suppression will have little impact beyond the communities in which these geniuses live and work. If there is no way for them to get the word out, others cannot benefit from their inventions and ideas.

So how can survivors get the word out about new alternative energy devices, drugs and other such vital information? Through multi-mode, multi-band, HF DX communication.

Of course there are naysayers who will insist that such ideas are the Pollyannaish pipe dreams of hopeful fools, which of course is the easiest thing for an arrogant pessimist.

In order to cast dark shadows on any future hope, pessimists enjoy spouting ill-considered assumptions with reckless bravado and to spot them is easy. They often start a negative rant with something like, "it seems logical to me..." or "everyone knows that..."

What do these pessimists offer in the end? A dark fate filled with futility and death due to the absence of any order or governmental structure. They believe that we'll revert to a Hollywoodesque Mad Max world where the good are certain to die. The only choices they offer us are about when and how we die.

Come on folks. As the singer Peggy Lee once sung, "Is that all there is?" Is it just negative pronouncements that we must throw away our lives by paying homage to arrogance and stupidity? If this is your choice, then good luck with it. As for the rest of us, how do we get the word out?

Getting the Word Out

Previously the point was made that line-of-sight VHF/UHF will get us through the end of this civilization but that only amateur HF offers us a clean slate to help lay a new foundation for the next civilization. For HAMs the may sound odd, so let's take a deeper look at what this means.

As our present civilization fails, so will the grid, the internet, municipal water and waste systems, things which we now take for granted. The greatest number of available two-way radios during this time will be bubble pack consumer FRS/GMRS walkie-talkies. This channelized service is in the upper UHF spectrum region and above the amateur UHF on the 70 cm band.

There will also be the amateur VHF/UHF handie-talkie (HT) and mobile transceivers. HTs, of the two, will have the greatest numbers available. This is due to the large number of HAMs with a technician class license. Generally, the two-way radios that newbie HAMs tend to buy are 5 W HT multi-band HTs for working local radio club repeaters. They will either become frustrated then and quit, or they'll step up to a more powerful 50 W mobile. From there it's on to HF.

Before the days of line-of-sight amateur VHF/UHF transceivers, HAMs could only work the HF spectrum region and had to demonstrate proficiency with Morse code.

Today, new HAMs who are ready to step up to HF, must upgrade their technician license to a general class license or higher. Then, there is the expense of purchasing an HF multi-band, multi-mode mobile, portable, or base station transceiver, which is considerable. One could think of it this way. With amateur VHF/UHF, you're in the hundreds of dollars. With amateur HF, you're in the thousands of dollars unless you can find a used bargain.

Once HAMs do step up to HF, the first mode they'll typically work will be phone. Then, they'll begin exploring the other many modes available and here is where many choose to learn Morse code so they can work long distance (DX) contacts using the CW mode.

As far as the channelized Citizens Band (CB) service, ninety percent or more of the available CB transceivers will be the basic 4 W AM phone mode type with a very limited range of just a few miles. The other ten percent of available CBs are Single Side Band (SSB) CB variants. Capable of DX contacts at 12 W, they operate in the HF spectrum region on the 11m band.

Limited in terms of power and restricted to phone mode use, it is difficult to say how effective SSB CBs will be with DX communications during a global tribulation. But, they will be helpful; so leave no stone unturned in finding these.

What is the upshot of all this? It's in the sheer numbers, as in the number of consumer FRS/GMRS and VHF/UHF amateur line-of-sight transceivers vs. the number of DX capable HF amateur and SSB CB transceivers.

This is why line-of-sight transceivers of all service types will become communication workhorses for survivors. They will have to deal with multiple community and local security issues as the infrastructure that shapes life as we know it, fails.

Will HF transceivers play an important role during this time? Yes, as survival communities stabilize, they can then spare the time and manpower to seek DX contacts with other stable communities for sharing survival knowledge over the air. A communication protocol I call "Knowledge Exchange."

What is to be gained with this protocol?

Knowledge will not only save lives; it will also help those who survive the worst of a global tribulation to thrive on "the backside," a future time, when those who make it, will once again see blue skies and taste sweet waters.

Forging DX Relationships

The goal of knowledge exchange contacts is to forge on-the-air relationships with trustworthy survivors for the sharing of survival knowledge. While this will occur wherever survivors can connect with two-way radios, please be aware of the following three concerns:

- **Proximity Threat:** Only in the most optimal conditions, can the range of line-of-sight phone mode services carry as far as fifty miles. For this reason, proximity to the community will always be a concern.

- **Filtering Out Impostors:** Regardless of the service or frequency, with phone mode contacts anyone can throw a few switches and turn a few knobs to go on the air, as impostors will seek ways to penetrate defenses. On the other hand, the majority of impostors will not have access to or understand how to operate an HF transceiver in CW (Morse code) mode.

- **Multiple Modes:** Modern HF transceivers are designed to operate in multiple modes. Some modes will be good for making initial contacts such as phone and CW, but other modes such as digital, facsimile and slow scan TV will be ideal for transmitting images of schematics, formulas and so forth over the air.

When forging new relationships for knowledge exchange contacts, the further away they are the better, with regards to security proximity concerns. On the other hand, the 50 mile line-of-sight limitation makes security proximity a constant concern.

Furthermore, by making Morse code CW contacts in the HF spectrum region, you have a safer way to determine if someone is being honest with you than phone mode. There are no guarantees, but most HAMs are good people.

Ergo, the advantage of making Morse code contacts is that you reduce the number of impostor contacts you need to filter out, as CW is not something an imposer can easily do by throwing a few simple switches or turning knobs. It requires training, diligence and the proper equipment.

Granted, you could find yourself communicating via CW with a HAM who has a gun to his or her head, but in the absence of proximity concerns, how likely is that to happen? Again, there are no absolutes but there are ways to shift the odds in your favor and absent a proximity threat, the only issue of concern is the exchange of useful knowledge.

What is useful? Not pleas for help as there will be many of those. Rather, you will seek contacts with others who have useful knowledge to exchange. This is why you make Morse code proficiency a high priority skill for your survival community.

Morse Code

Morse code dates back to the days before radio when it was used in telegraph offices to send messages. It was invented in 1836 by Samuel F. B. Morse, the inventor of the telegraph, and it is an internationally recognized method today of transmitting text information as codes via on and off tones or clicks, typically described as dots and dashes.

Morse code was first used with wireless radios in 1901 when Guglielmo Marconi sent the first Morse code transmission across the Atlantic from England to Canada. He used what is known as a spark gap transmitter. These were very primitive and inefficient transmitters.

After WW I, the use of vacuum tubes allowed for more sophisticated transmitters using the continuous wave (CW) mode and these second generation CW transmitters were far more efficient than first generation spark gap transmitters. They could transmit clean signals on specific frequencies as opposed to splattering across a band and with an efficient use of transmit power.

Consequently, CW remains the most efficient of all modes in use today for sending Morse code messages across vast distances. It also sounds different from first generation spark gap transmitters and so HAM operators presently use the term "dits" for dots and "dahs" for dashes to describe how Morse code sounds with CW mode.

Now let's give this a practical application with the one thing that is more important than anything else. When it comes to creating an emergency radio system, it is the ability to transmit and receive Morse code.

Morse Code Proficiency

Morse code proficiency boils down to how fast can you accurately send and receive Morse code. This ability is essential at all levels and especially within families.

Until the complete abolition of the Morse code testing for FCC HAM licenses in 2005, the proficiency testing depended on the license class. Legacy Novice Class and Advanced Class licenses do continue. However, as of April 2000 the FCC no longer issues new licenses for these legacy classes.

The Morse code requirements once used by the FCC to test proficiency are very practical and useful in terms of defining the requirements for the three levels of your community communications strategy.

Current Class	Legacy Class	Legacy Speed	Community Level
Technician	Novice	5 WPM	1 – Emergency System Radios
General	General	15 WPM	2 – Field Radios and Repeaters
Extra	Advanced	20 WPM	3 – Operations and RFE Broadcasting

What caused the FCC to remove the Morse code requirement was the advent of cell phones. As cell phones became more popular, the number of annual applications for HAM radio licenses began to drop. The result is that the entrance of new generations of HAMS into amateur radio started to become progressively smaller. To encourage new HAMs to enter amateur radio, the FCC dropped the Morse code requirements.

One could imagine that this FCC change in testing may have caused a massive reduction in the number of HAMs working Morse code, but the opposite is true. More HAMs in America are working Morse code now, more than when proficiency testing was a requirement.

Why is this? People are people and at some point you get tired of someone telling you to do something or else. That takes choice off the table, and the fun of doing it as well. However, now that Morse code is a personal choice and not a Federal requirement, more HAMs have chosen to learn it than ever before.

Another benefit is the nature of the communications traffic. With phone mode, professionalism is a challenge and depending on the service, there can be some degree of unprofessional conduct.

However, Morse code is one place where professionalism reigns supreme. There is no arrogant disdain or showdown. Rather, newbies find a welcoming environment where they are treated with kindness, patience, respect, and compassion.

Morse Code in a Global Tribulation

With the catastrophic failure of various elements of the nation's infrastructure at the outset of a global tribulation, smartphones will become useless as communication devices and will be reduced to music and video applications. While those functions will be useful, the fact remains that radio communication will become more essential than ever before, during this time.

It makes sense to require universal proficiency for certain skills like microphone technique. However, the lessons learned with FCC license testing for Morse code is different and presents the following three fundamental leadership concerns:

- **Choice:** There will be members of your community who will simply not want to learn Morse code. Respect their choice and never proselytize the skill. If by virtue of their interactions with members of the community who are proficient in Morse code they take a future interest, be welcoming and helpful.

- **Attrition:** Ask any Morse code military instructor about teaching Morse code and one of the first things they'll talk about is attrition as an immutable constant and it can happen at any time. Whether it is shortly after the student begins or after a student completes the instruction, they can choose to walk away.

- **Motivations:** The key to mitigating the loss of attrition is to carefully evaluate community members regarding their motivations for learning Morse code. Without a doubt, some will leap at the opportunity to put down a shovel, so they can relax in a classroom. Do not waste time or resource on them with Pollyannaish thinking.

With these concerns in mind, the adverse conditions of a global tribulation require an altogether different approach. You need to inspire – not require.

Inspiring an Interest in Morse Code

Today, there are a multitude of Morse code learning resources with computer programs, smartphone applications, CD, and DVD that we'll discuss later on. But for now, let's go back in time to the very early days before radio was invented and before there were any code learning resources. What were the learning resources in those early days? Trial and error, but that was the nature of life at that time.

Will the same motivation work with today's youths and young adults. Not likely, because life was more of a struggle for the early telegraphers. There was no sense of entitlement as there is today and people could handle themselves better than today's young.

When forming your survival community, you'll have many undisciplined, uncommitted, young people and this is unavoidable.

Ergo, there will be opportunists attracted to the classroom comfort of Morse code training because they have a very narrow capacity for hardship and a sense of entitlement.

Is this to say that you should exclude the self-interested from Morse code training? Absolutely not, you need to find a different way to motivate them when possible.

What is that different way? The early wire and wireless telegraphers were communicators who relayed messages between senders and receivers. During a global tribulation the same need will apply, but there is an even more important reason.

Communicators and Facilitators

Rather than seeing themselves as communicators, which is admittedly a boring role, you need to inspire the young to become facilitators as well as communicators. This is what's new for the years ahead.

As was pointed out earlier, Morse code is a universal worldwide language. Given that we can translate web pages from one language to another with sophisticated web browsers, that may seem to be a frivolous advantage, but when the internet fails, the only universal worldwide language left standing will be Morse code.

If someone says give me proof of that, you have perfect proof right in this book in Appendix B – RMS Titanic Radio Log. The Titanic struck the iceberg at 11:40 PM on April 14, 1912. The Titanic sank 2 hours and 40 minutes later at 2:20 AM.

There is a general assumption that the Marconi radio operators on the Titanic only sent a few messages and then went off the air because of films like Titanic (1997). Producer James Cameron only showed a few brief moments of what occurred in the Titanic's radio room that tragic night.

Unfortunately, most of the radio room scenes shot for the film wound up on the cutting room floor, but even if they had been included in the final cut of the film, these deleted scenes would only have offered a little more about how the British steamship the SS Californian saw the Titanic's signal flares but ignored them.

In terms of Morse code as a universal worldwide language, the sinking of the Titanic was a turning point in the history of radio. Although the SS Californian ignored Titanic's distress calls, ships from many nations across the Atlantic responded. The ten ships that communicated with the Titanic that night were the: Asian, Baltic, Burma, Caronia, Carpathia, Frankfurt, Mount Temple, Olympic, Virginian and Ypiranga. An eleventh station, the land based station, Cape Race, communicated with the Titanic as well.

What happened that night was that Morse code operators set aside their mundane communicator tasks of sending routine reports and passenger messages and began doing all in their power to facilitate the rescue of Titanic's passengers. Even though wireless was very new then, many ship's captains across the Atlantic still raced to Titanic's aid, but it was the Carpathia that reached them first and only because they could get there sooner.

It is very rare to find the entire radio log for that evening in print and Appendix B has it all. From just before Titanic struck the iceberg to the moment that the ship went off the air.

The point of Appendix B is to help you teach the youth of your community that their understanding and proficiency in Morse code is not only a worldwide language of survival, it is also a gateway to a magnificent future of knowledge exchanges and cooperation between facilitators over the air.

As facilitators, the youth of your community can use their proficiency with Morse code to facilitate vital collaborative work between the subject matter experts in their community and those of distant communities over the air.

Instead of working with in and out baskets like the early Western Union telegraphers, they will be working with knowledgeable people in your community who need to collaborate with peers elsewhere in the world to mutual advantage.

They will work as vital facilitators for your community by sharing ideas and inspiring new solutions with others. They'll perform an essential role in pulling together pieces of a puzzle until all concerned can see a total and complete solution that helps survivors and saves lives.

Remember, you need to inspire – not require. Do this and you will build interest in learning Morse code more quickly.

Learning Morse Code

Today there are wide variety of teaching tools such as computer software, smartphone apps and audio CDs. While there are a variety of instruction methods, the two that are presently used are the Farnsworth method, named for Donald R. "Russ" Farnsworth, and the Koch method which came along later and is named after German psychologist Ludwig Koch.

The principal differences between the two as they are taught today are:

- **Farnsworth:** Individual characters are sent at a high speed, but spaced out from one another. Students learn to recognize both characters and words. The first and still the most popular method for many seasoned HAMs, the Farnsworth method is a great way to learn what could be described as "conversational Morse code."

- **Koch:** The focus is on learning letters at high speed. Students learn small groups of characters, usually two at a time. If you are sending encrypted code messages such as is done by military **telegraphers, the** Koch method is preferable to Farnsworth. This is because coded messages use a series of groups of random letters and numbers and are often transmitted by machine or what is called "flat code" because it has no human personality.

Both methods have their unique advantages and teaching styles, and it is not uncommon to see HAMs learning Morse code using a combination of both teaching methods.

The bottom line is that the final choice of which method or method mix needs to be guided by an experienced HAM in your community who is proficient with Morse code at speeds of 15 WPM or higher. The role of this HAM needs to be that of an Elmer, an amateur radio term for tutor or mentor.

Culture of Communication

Remember, you need to inspire – not require. In this case that means that you want a more organic, viral process for propagating Morse code proficiency throughout your community to create a culture of communication. The purpose for learning Morse code will be very different during a tribulation than those which came before.

Contemporary motivations for learning Morse code can be categorized as: economic, conditional, and personal interest.

In the early days of the telegraph system, economic considerations were a primary motive. Telegraphers did not learn Morse code using modern teaching methods and tools. Back then, the traditional method was to struggle through all the codes at a slow speed; then with practice, achieve higher speeds over time and their motive was economic.

Instead of the sweat and misery of pushing a plow, digging ditches, baling hay, laying railroad track, etc., young men could take a job as a telegrapher, a true, clean, and comfortable desk job with a nice chair. They could stay dry in winter and be relatively comfortable in the summer. Best of all, these telegraphy jobs paid well, even though they pretty much had to learn Morse code on their own.

According to historical accounts, novice telegraphy students took about a week to learn how to send and receive messages rapidly and efficiently with Morse code. Then they could start work as a telegrapher for Western Union and other telegraph services

In a similar vein there are the military services, and here we see a conditional motivation. Like the telegraph operators, the role of military radio operators is to relay messages and they are trained and receive a military classification like the US Army Military Occupational Specialty (MOS). Consequently, the motivation for the radio operator is to faithfully train for and serve in an MOS. It may well be a condition that motivated their joining the military..

The third motivation is personal interest. For HAMs, experimenters and hobbyists, this motivation is neither conditional nor is it economic. It is about choosing to pursue an interest for personal reasons.

What all three motivations, whether economic, conditional, or personal interest, have in common is that they apply or appeal to the few and not the many.

During a global tribulation, motivations that only appeal to a few will create a burdensome layer of dependency in your survival community upon those who choose to serve the many, assuming there are enough of the few to be successful.

Keep in mind the power of Morse code. It is not only the most efficient way to send messages over wire and over the air; you can also send it with flashlights, drums, or banging on a wall. You can even use it to communicate with the survivors of a cave in.

This is why a completely new approach to propagating Morse code proficiency throughout your community requires going "outside the box" as they say in the Silicon Valley when creating a new motivation for learning Morse code.

To do this, you begin by engaging the youth in your community. Give them free reign with the support of a HAM Elmer who is highly proficient in Morse code to experiment as a team. Note the term support as opposed to instruction. To be inspired, the youth must lead.

Do not be surprised to see HAMs shake their heads at the notion of giving the youth community free reign to create a culture of communication based on Morse code proficiency. Then again, they will typically be senior citizens and their motivation for learning Morse code was economic, conditional, or personal interest – whichever their frame of reference was.

When faced with the resistance of contemporary motivations, do not ignore the need for the HAM or Elmer to be inspired as well. The last thing you want is for someone who is supporting your youth in this effort to say "I would have done it differently." Statements like these are the seeds of defeat; so do not let them take root.

Rather, you should explain that under the new conditions there are different motivations that will create viral support for Morse code proficiency throughout the community.

What do you do with intransigent HAMs who continue to believe that your Morse code proficiency goals are bound to fail because they are too set in their ways to think outside the box and be team players for survival? Put them in roles for which they are qualified, but minimize their contact with the youth of the community until they've adapted to the new circumstances of a global tribulation. This would include assigning a full time apprentice.

As for those who will learn Morse code, are there preferences? Yes, while anyone with an interest and basic ability to work Morse code can learn it, the younger they are, the faster they will pick it up. How old should they be? Those who are in late pre-puberty will make the best learners, but you'll also work with teens and adults as well. An interesting thing was learned about late pre-puberty age children.

Years ago the government funded a program for creating aquaponics systems (systems that combine aquaculture and hydroponics) for colonies on planets such as Mars. What the program developers found to their amazement was that children of pre-puberty age were hands down better at increasing crop yields than any adult professional with a PhD on the team.

Unlike adults who tend to view the world through what they've experienced and learned, these children saw the plants as-is and behaved without preconceived notions or pet theories.

Through a process of experimentation, observation and collaboration, they worked their own thought processes to make the impossible possible. Use this insight to your advantage and amazing things can happen. When children become involved as significant creators, they will also create interest within their circles of family and friends as well.

Morse Code Learning Tools

What is the best way for you as a leader to teach Morse code? You don't. Never do it. Never!

Rather, you buy every teaching tool for learning Morse code and give them to the kids to sort out on their own. They may use a little here and a little there, but you're still only talking about a hundred dollars or so.

What will cost you more are the physical tools, of which there are primarily three:

- **Key:** A key is a simple mechanical device used by the operator to compose the dits and dahs of a Morse code message and there are several types. The oldest and simplest are called straight keys or telegraph keys. Your community will use them with Level 1 emergency radio kits.

- **Keyer:** The difference between a key and a keyer is that keyers require additional electronics to work, hence the term "keyers." There are several types of keyers but the one that is the most popular by far is the dual-paddle Iambic keyer. Your community will use them and straight keys with your Level 2 and 3 transceivers.

- **Oscillator:** A typical feature found in HF transceivers that support CW mode is something called a "sidetone." A sidetone allows operators to hear what they are sending, but you should never use a sidetone for training as this is a very inefficient way to use an expensive transceiver. Rather, you use a simple stand-alone device called an "oscillator" connected to the key or keyer to duplicate the sidetone function.

An inexpensive and versatile Morse code teaching tool is the MFJ-557 code oscillator with straight key. It is a combined key and oscillator that runs on a 9 V battery and can also be powered with a 12 VDC battery as well. Its well-thought-out design for students has volume control, earphone jack, external speaker support, and user-adjustable tones.

When you equip your team with the necessary training tools and methods, you also need to give them basic starting parameters. These are:

- **No Visual Aids:** Morse code uses hearing and tactile senses – not vision. If you use visual aids like letter charts, you'll only create horrible learning impairments.

- **Train With a Straight Key:** The advantage of a straight key is that the operator has full control which in turn helps them to develop what is called a "fist." Their way of sending code will be just as recognizable as the code itself. Keyers, on the other hand, create a less recognizable fist due to the automation.

- **Do Not Start Slow:** In the old days when newbies were working to get their Novice class, they set their sights on the 5 WPM required by the FCC. Then they would work their way up. Very bad idea.

  You want to start with a higher speed such as 15 WPM or even better 20 WPM. Sure, students will have a lot of errors by starting at a high speed, but the truth is; it is easier to decrease the error rate than it is to increase speed.

There are other useful tips, but the fewer starting parameters you begin with the better because your team of youngsters will learn these and find others on their own as they proceed. Therefore, give the process a wide palette of tools, techniques, support and as much free reign as possible.

Then there is one more thing. You need to build the first Morse code team and that requires planning and patience.

Building Your Morse Code Team

When assembling your first team, the very last, last step is for them to learn Morse code.

The first step is to assemble the team. Initially, a team of committed adults is required. An experienced teacher and at least two local HAMs, who must be proficient in Morse code with both a straight key and a dual-paddle Iambic keyer.

The ideal teacher would know absolutely nothing about Morse code but understand what motivates kids and how to recognize true talent and commitment.

This teacher needs to have a genuine interest in learning Morse code and how to operate every transceiver in your community. This is a good role for a woman who will have the full support of mothers. Ideally, then, the time will mainly be spent on learning and not wasted on resolving disputes. When you find that special teacher, assign to her the title and responsibilities of the Training and Support Coordinator.

Her first task will be to develop a screening questionnaire that uses a point system to rank Morse code candidates. This will not only help to filter out but also to find the kids with a gift and a committed interest in the technology; what HAMs call "the bite."

An example screening questionnaire could contain the following criteria where each item can be scored from zero to five for a final tally. Sample items include:

- **Outdoor Activities and Sports:** You'll need field radio operators for your field teams who are physically fit and ready to go into the field.

- **Scouting and Civil Air Patrol:** The number of badges is more important than the types and the more merit badges the better.

- **Multi-lingual:** Learning Morse code is easier for those who are multi-lingual. English should be the primary language.

- **Science and Engineering:** If you find a kid who has competed in two or more science fairs, this is a solid 5. Video games are not science and engineering.

- **Blind or Disabled:** Remember, Morse code is auditory and tactile. It does not require you see something or to walk towards something. If a child is blind or disabled but physically capable of being a competent telegrapher, then whether they're sitting in a chair or a wheelchair makes no difference. If they have a genuine interest, regardless of inability to participate in the field, they are a solid 5.

- **Musicians and Artists:** A not-so-secret secret in the Silicon Valley is that musicians can become amazing software programmers and often do. The same holds true for Morse code.

- **Competitive Achievers:** Kids with an interest in business achievement such as the DECA program or who enjoy competing in speech teams or chess teams for example are a solid 5. These are your go-getters.

This is just a partial list of sample criteria and your Training and Support Coordinator will need to carefully design this questionnaire with the help of your HAMs and if possible, skill-related professionals in the community.

What you want to achieve is the lowest possible attrition rate possible while finding different kids in the community who are suited to one of the following roles:

- **Training Assistants:** Your Morse code kids will be your first training team. Later, you can expand to other roles based on whatever criteria works best for your technical team and HAMs. The more kids you can involve as training assistants, the better.

- **Level 1 Emergency Operators:** These kids will be trained with a straight key only; a proficiency of 5 WPM is adequate. This will be your largest pool of operators.

- **Level 2 Field Operators:** These kids need to be proficient with keys and keyers equally and have an effective rate of at least 15 WPM.

- **Level 3 Base Station Operators:** Ideally, these kids will become 24/7 apprentices assigned to a HAM. Your HAMs will be old and there will not be a lot of modern medicine for them during a global tribulation. This is why you need to pair them with young 24/7 apprentices who will be equally proficient with keys and keyers and able to achieve effective rates of 20 WPM or higher with a keyer.

These are all general guidelines that follow conventional standards, but there will be those kids who could come in last on the questionnaire and yet, become your fastest *telegraphers*. *The last thing after reviewing a questionnaire is to trust your gut instincts before you decide.* Again, you need to inspire – not require – but how is that done?

Inspiring Your Youth

Today's Generation Z is also called the iGeneration and for good reason. They grew up with handheld devices, the Internet, and Wi-Fi. Hand them a smartphone and what you see is amazing. While Baby Boomers fumble with hunt and peck failures, these children see these devices as an extension of their own bodies and minds. They just do it!

Yet, show them a straight key and they'll probably recoil from its obvious simplicity and view it as hopelessly inefficient when compared with their smartphone.

You can quickly dispel this with a five stage process such as the following example:

- Watch the Titanic Movie
- Perform the Leno Test

- Build and test a DIY Spark Gap Transmitter
- Read the Titanic Radio Log Aloud
- Inspire Them

The event begins with the Training and Support Coordinator gathering the candidates together in a comfortable setting to view the Titanic movie on a widescreen if possible. Tell the kids there will be treats and a discussion after the viewing.

Once they've viewed the film, get them to talk about the decisions people made in the movie about survival. Make it clear before the movie, that the discussion will not include the romantic story line. Make that clear – no survival choices, no treats.

At the end of the discussion, the Training and Support Coordinator explains that they will meet the community's HAMs for an exciting experiment. The group should then move into a classroom setting and there be formally introduced to your HAMs.

Allow time for this encounter as your HAMs need to get a good feel for the personalities and abilities of the group. Then, you move to the Leno Test.

On the May 13, 2005 episode of The Tonight Show, host Jay Leno conducted an on-air speed contest to a large audience to see which was faster. Morse code telegraphers or smartphone users who are fast at texting (SMS)

The HAMs were seasoned Morse code competitors, Chip Margelli (K7JA) for sending and Ken Miller (K6CTW) for receiving. The two smartphone Millennials were text-messaging world champions Ben Cook and his friend Jason.

Nearly everyone in the audience assumed the Millennials would smoke the telegraphers, but that was not how it worked out. When the two teams were given the same message, the result was the telegraphers easily beat the smartphone Millennials, to everyone's surprise.

For HAMs what was particularly amusing was that Chip Margelli (K7JA) had an effective rate of 29 WPM using a dual-paddle Iambic keyer. How fast was that when compared with the smartphone Millennials? Here is the kicker.

When Leno asked about the speed, Chip Margelli said he could have finished quicker but used the time to send the entire message a second time while the smartphone Millennials were still texting the same message for the first time.

Recreate this Leno test with your kids and HAMs if you can; you'll surprise them, especially the kids, so no spoilers. Then after that sinks in, tell the kids that Morse code is now the new Tribulation Twitter solution. They'll get it.

Low Power AM Band Spark Gap Transmitter

- Two Wood Screws
- Small Block of Wood
- 9 Volt Battery
- Red & Black #24 Gauge Wire
- Straight Telegraph Key
- Electronic Fly Swatter Circuit

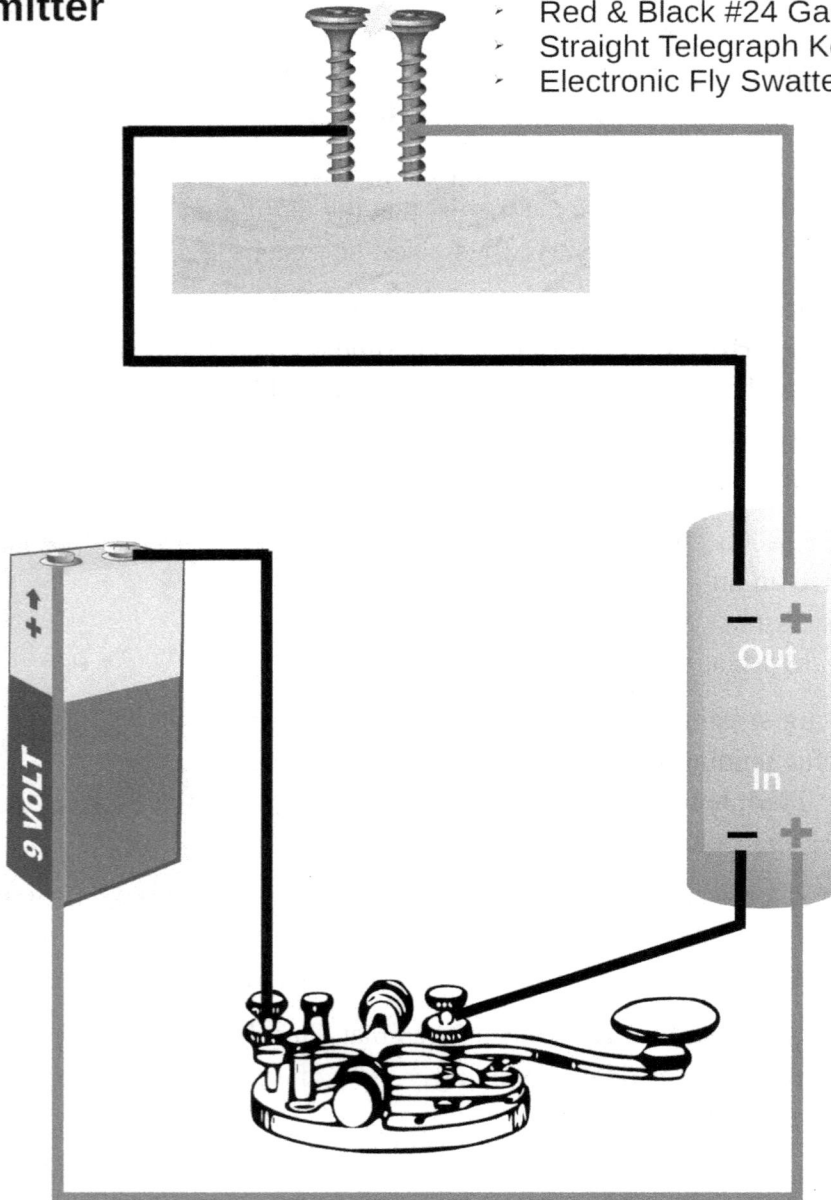

9 VOLT

Out

In

Next you organize a DIY assembly project to build a spark gap transmitter. It was a state-of-the-art Marconi spark gap transmitter that was installed on the Titanic and building a small example of one can be done with a cheap electronic fly swatter and a few components as illustrated above.

Once the kids have assembled their micro-watt DIY spark gap transmitter they will test them with two, three letter words:

- **SOS:** The standard international distress call. It has no real meaning, though some like to say "save our ship."

- **CQD:** First used by the British and American ships, CQD stands for "attention all stations distress." This distress code was replaced by the SOS distress call.

The reason you want them to test their spark gap transmitters with these two distress calls, is that it was the Titanic disaster that made SOS the new international distress call.

At this point the kids will have seen and talked about the movie, built a DIY spark gap transmitter, and tested them with these two distress codes. Now you go for effect.

Give each kid a candle and a copy of the Titanic radio log found in Appendix B. Once everyone is settled, explain that the kids will take turns reading each log entry until the entire log has been read.

When they are not reading, they hold their candles low to signify receive mode.

Then before they read each entry, they raise their candle to signify transmit mode.

After explaining this, the HAMs leave the room and the Training and Support Coordinator turns off the room lights and the reading begins. As that takes place, you are notified to come and speak to the group.

After the reading is finished, the Training and Support Coordinator announces you and you enter the room. That's when you inspire them to become future communicators and facilitators.

Inspire them with purpose.

Inspire them with love.

Believe in them and the rest will just happen.

23

News Gathering and Broadcasting

Broadcasting over any consumer or amateur radio frequency today is expressly prohibited by law. Try it, and FCC enforcement officers will find their way to your door, and you will have to endure confiscation, severe fines, and possible imprisonment.

However, during a global tribulation, when FCC enforcement is no longer possible and commercial stations are going off the air, things will change. Enforcement will be local and there will be an absence of operational commercial broadcasters. Therefore, do not assume that survivors will not be seeking high-quality survival broadcasting which will be absent. The reverse will be true. In such times, it will be more necessary than ever before.

Addressing the need, this is why you'll want to slowly and carefully build a local Radio Free Earth Network that begins with reliable news reporting based on a simple operational strategy: verified intelligence in – reliable news out.

To illustrate agenda-free reliable reporting, let's assume that Homeland Security forces in your town have received 50 new Mine-Resistant Ambush Protected (MRAP) vehicles that residents have observed, and these vehicles are staged at a local Walmart which is used as a Forward Operating Base (FOB). Let's look at two examples of how the story is reported, one with and one without an agenda.

Agenda-free	Agenda-driven
"We have received multiple reports from reliable local sources that the fifty armored vehicles, known as MRAPs for Mine-Resistant Ambush Protected, arrived yesterday at the local Homeland Security base at the Walmart store at 6th and Main in Bakerville. RFE will attempt to interview the local Homeland Security commander about this new development and will report the results as soon as possible."	"Fifty armored vehicles, known MRAPs for Mine-Resistant Ambush Protected, have been received by the local Homeland Security base at the Walmart store at 6th and Main in Bakerville. Residents have expressed fear that these heavily armored vehicles are destined for early morning, warrantless raids, and arrests such as those conducted by U.S. armed forces in Iraq and Afghanistan."

While both examples contain factual reporting, the key thing to remember is that by this time we are living in a Martial Law environment. Ergo, the last thing you want to see coming down the street is a Bradley Fighting Vehicle or an MRAP because a hothead goes off half-cocked because of a provocative broadcast.

Radio Free Earth Affiliates

When it comes to affiliates, they are not your friends. They are affiliates, and if they go off half-cocked, there can be no second chances so make no exceptions!

Look at the front cover of this book which features a photo of a Radio Free Earth Network patch with the motto "Integrity, Objectivity, Service." Let's take a closer look at this.

- **Integrity:** In the first Voice of America broadcast in 1942, listeners heard, "The news may be good or bad for us – We will always tell you the truth." Use this level of integrity to shape your broadcasting model and your efforts will earn similar respect.

- **Objectivity:** When you compare the Radio Free Earth motto with what we're seeing in the broadcast news media today, it fails because what we presently have is agenda-driven news entertainment and propaganda. There is a considerable amount of agenda-driven news already, and during a global tribulation there will be ten times more.

- **Service:** You must fully embrace a service-to-others approach for ALL survivors when building your local Radio Free Earth Network. If there is the remotest hint of self-service, disinformation, misinformation, or manipulation, you'll create enemies and risk being seen as an elitist disinformation operative.

All who are fully committed to the Radio Free Earth motto, Integrity, Objectivity, Service, are welcome to identify their local network as an affiliate. Every member of their local network must agree to adhere to the following requirements faithfully:

- **Location Identifier:** Whether in person or on the air, you must use a general location identifier. For example, with a network located in Arizona, you could identify it as Radio Free Earth Flagstaff, or Radio Free Earth Tucson. You should never be specific about the actual physical location of your survival community.

- **Network Badge:** When investigating news stories in public such as at town meetings, man on the street interviews, and so forth, wear a recognizable network badge such as the one featured on the front cover this book.

- **Professionalism:** You must always conduct yourself professionally with no foul language, while demonstrating courtesy, and respect for all. Without this, you cannot earn the respect of others for being fair-minded and even-handed. Only in this way can you be valued by the public at large.

The point of creating a local Radio Free Earth Network is to provide reliable, agenda-free news reports. If you intend to pursue an agenda, such as promoting a religious faith through a church, do that as a completely separate organization.

It is perfectly acceptable to announce at the beginning and end of a news broadcast that another organization supports your news efforts. Using the example above as a reference, you could announce that support like this, "Radio Free Earth Tucson is made possible by the Tucson Church of Perpetual Genesis. To learn more about our church, please tune to (channel/frequency x) at (time) each day for our expanded programming."

It is likewise appropriate to add an announcement to promote your independent Radio Free Earth network in this manner, "The Tucson Church of Perpetual Genesis is a proud supporter of Radio Free Earth Tucson. To learn more about this local news network, please tune to (frequency) at (time) daily for reliable news reporting."

Remember, the minute you use Radio Free Earth to promote or proselytize a political or religious point of view, those who subscribe to those views will enjoy your programming. The rest will see you as promoting, possibly untrustworthy, and entertainment.

Therefore, if you are going to pursue an agenda-driven broadcasting program, please respect the author's wish that you NEVER identify yourself or your organization as a Radio Free Earth Network affiliate. Likewise, the author asks that you NEVER claim to be associated in any way with any other Radio Free Earth Network affiliate.

Intelligence Gathering

There are many ways to gather news, and each must use the same operational strategy: verified intelligence in – reliable news out. Why intelligence as opposed to the more familiar term, investigative journalism?

In the last few decades, politician appointees to the FCC have followed a methodical policy of eliminating anti-concentration policies in favor of the ruling elites. The goal of this policy reversal is clear, to make it financially impossible to support true investigative journalism as seen in years past. The result of this policy reversal is the current plague of "Fake News" and public disgust. No wonder journalists are now proclaiming that the age of investigative journalism has largely passed. There is nothing to be gained by hitching your wagon to the dead horse of a bygone age.

On the other hand, intelligence gathering is a more straightforward and reliable process, and there are three things your local Radio Free Earth network affiliates can do when gathering intelligence for community operations and news reporting.

Two were previously discussed at length; though a summary is helpful. One is the Five W's and How and the other is the The Five Power Listening Skills.

Professional journalists use the Five W's and How to gather the facts needed to assess a situation and then to create a complete news account. Here is what they learn:

- **Who** was involved?
- **What** happened?
- **When** did it take place?
- **Where** did it take place?
- **Why** did that happen?
- **How** did it happen?

Answers to each of these six questions must be factual and in detail. "Yes" or "no" answers are wholly unacceptable. Keep your questions simple, polite, and professional. Do this, and you will firmly establish the integrity of whatever news you report.

Also, using power listening skills will help you encourage confidence in your ability to be objective. The five power listening skills are:

- **No Clipping:** Do not interrupt before a person can complete what they are saying.
- **No Anticipating:** Do not think about how to respond to someone as they are speaking.
- **Listen Thoughtfully:** Listen thoughtfully to someone with an attentive and calm demeanor as you take note of what they're saying.
- **Pausing for Focus:** Once a speaker finishes what they have to say, continue to maintain direct eye contact, but remain silent for a few seconds before you answer.
- **Ask Confirming Questions:** Of all five listening skills, this is the most powerful of all. Engage the speaker with questions and thereby find clarity.

The third thing you can always do is use reliable sources. For example, you may have a long haul trucker that passes through your area and can offer an interesting account of events from other areas. Here is where you must decide whether you are an entertainer or an investigator; so what is the difference?

- **Self-serving Entertainers:** Entertainers always disavow responsibility for the authenticity and veracity of the information they air. They tell their audiences to decide if the facts presented conform to the truth because their job is just to make it available. By doing this, they are then free to focus on finding sensational images and stories that will generate audience traffic without the burden of vetting and research. They just spew it.

- **Service-to-Others Investigators:** During a global tribulation, authenticity and veracity will not simply be goals. They will be a true expression of objective service to others because you are a neutral party. What does neutral mean? If you are faith-based organization that uses news reporting to attract an audience to your network so you can proselytize your religious views, you're a self-serving entertainer; no matter what standards you apply to your news reporting.

Can there be a successful entertainer during a global tribulation? Yes, until someone else steps up with reliable news reporting based on a true service-to-others commitment. When that happens, you'll fade away as an anachronism.

You can create effective intelligence gathering for news by building a network of reliable two-way radio "stringers." In the news reporting business, stringers are independent correspondents, and they are usually retained on a part-time basis to report on local events. During a global tribulation, you can expect difficulties in finding objective journalists to work with your local Radio Free Earth Network.

Whether you find your stringers personally or over the air, you must always vet them. This must be done either by direct interviews by your network affiliates or by reliable and trusted sources. You can listen to strangers or one-trick-ponies looking for their fifteen minutes of fame all you like, but the moment you air their intelligence without corroboration you become a self-serving "Fake News" entertainer.

Finding locals that are reliable, have two-way radios, and who are team players by nature, is the way to go. Getting started is easy. Organize them with what HAMs call "radio nets."

Radio Nets

Amateur radio nets, also referred to ham nets, are how HAMs meet over the air. These nets have the following attributes:

- **Scheduling:** Nets are scheduled events. They can be one-time events or recurring events that repeat daily or weekly, as a rule.

- **Frequencies:** When scheduling a net, the broadcast frequency is important. The convention is that they must not take place on high traffic frequencies such as a national calling frequency. Consequently, nets are typically organized on the low traffic frequencies to avoid interfering with other operators working the same frequency.

- **Net Control Station:** Unlike accessing chat rooms on the internet via software automation, radio nets are a purely hands-on, manual process. Someone starts the net which is essentially like starting a chat room and then others join in. With amateur radio this organizational role is performed by the Net Control Station.

- **Purpose:** Nets serve any number of communication needs such as relaying messages, training, testing equipment or discussing a topic of interest.

These nets will become very important for your technical team and level 3 HAMs during a global tribulation. This is especially true when they're testing equipment, such as new antennas or old two-way radios that have been salvaged or repaired.

More to the point, you can build a phenomenal team of local stringers with scheduled news-gathering nets which will be unique to the coming global tribulation.

Newsgathering Nets

Are radio nets limited to a specific service or band? No, because starting a net is a manual process on any frequency and for any two-way radio service. Nets offer a powerful way to connect with people. If one person is using a new high-tech radio and someone else is using a 30-year old relic, there are no compatibility issues. If both operators can send and receive on the same frequency, they've got a fully functional net.

To illustrate the concept, let's look at three possible categories of newsgathering nets that you can add to your community communications strategy:

- **Local Channelized News Nets:** The most numerous two-way radios that will be able to join a net will be CB and GMRS. Each net for a specific channelized service should be scheduled at different times since some locals may want to participate in both.

- **Local Amateur Emergency News Nets:** Here is where a cross-band repeater can give you a powerful advantage. Since there will be far fewer tuneable VHF/UHF amateur transceivers available for these nets than the channelized two-way radios, you'll have wider coverage. A good policy would be to restrict net participation.

- **Long Distance News Nets:** Both options in this category are in the MF, HF, and VHF spectrums. With frequencies assigned to amateur transceivers, the same restriction applies. They need to have a call sign.

Here is one area where SSB CB is an odd duck so to speak, because SSB CB is a shortwave service in 11m HF band and it is worked by unlicensed CB enthusiasts. A judgment call is required here because local SSB CB operators will often be farmers and other rural operators. Do you use them for local or long distance contacts?

Rather than trial and error to see how things go and then decide, a more methodical approach is available through the type of radio net you choose.

Radio Net Types

Earlier, we examined the role of local amateur HAM operators as the front line of your local Radio Free Earth network. They will operate the news-gathering nets with local citizens and other reliable sources to acquire and then relay vetted news reports to your news production team, and there are two types of nets:

- **Formal Nets:** Also referred to as "directed nets," each formal net session is managed by a single Net Control Station (NCS) who calls the net to order at the scheduled time. The NCS typically begins the net with announcements and then moves on to station logins. Logins can be organized according to a pre-assigned roster of allowed stations or with an open check-in by other stations.

- **Informal Nets:** An informal net may have a pre-assigned NCS, or the NCS becomes whichever station arrives first. These types of nets are ideal for technical discussions and for testing equipment, since the NCS controls who transmits and in which order.

What is required of all net participants is a working two-way voice radio, a basic understanding of the radio net procedures, and unreserved respect for the NCS's instructions. For news gathering, a formal net with a preassigned roster of allowed stations is ideal for certified affiliates and stringers.

Then local news gatherers, stringers, and affiliates can operate informal nets as the NCS for gathering news tips from local citizens and the travelers that pass through. Verification of these news tips is important. If you report an unverified tip by an anonymous disinformation operative or troll, it may cause a needless panic and your reputation will be stained. Trust in your network as a reliable source of news will diminish, and this may also create enemies.

Therefore, when you conduct nets, it is important to remember that your nets will be heard openly, since there is no encoding or encryption on public channels. While only a handful of participants may check into your nets, you should also assume that others will monitor your scheduled nets. They will, and in numbers greater than you may imagine.

Lurkers will listen. While most will appreciate your nets for their timeliness and the professionalism of the news traffic shared, some will undoubtedly listen for opportunities that they can exploit or manipulate with evil intent.

So, always conduct your nets professionally and the public at large will view them as a useful and valuable source of reliable survival news and information.

The next step is the broadcasting of that information as part of a broadcast program.

Channelized Broadcasting

As discussed previously, the largest number of two-way radios available during the global tribulation and beyond will be FRS/GMRS and CB. The advantage of these channelized services is that they will be useful for newsgathering nets and also a good place to begin broadcasting operations during emergency conditions when FCC enforcement is no longer possible, and commercial stations are off-air.

Whether you are broadcasting with FRS/GMRS or CB, the same transceiver considerations apply.

- **Intended Use:** Channelized and Amateur transceivers are not intended for use for extended broadcasting. Rather, think of them like the powered homeowner rated lawn care tools you store in your garage that are designed for a few hours of weekend use and with frequent ice tea breaks.

  Contrast them with the commercial-rated powered tools used by professional landscapers. They are more expensive because they are designed for daily and long-term operation and have a considerably longer service life.

- **Transceiver Duty Cycle:** The duty cycle is the percentage of time the transceiver should be in transmit mode as determined by its manufacturer. For example, a duty cycle of 25% at full power translates to 15 minutes of continuous transmitting each hour. Exceed the time allowed for the rated duty cycle, and you will burn out the transceiver.

  The duty cycle can also depend on the mode. Using phone mode (for example for broadcasting) at low power may give you a duty cycle of 25% with a channelized transceiver. However, an amateur transceiver working in low-power CW mode can have a power duty cycle of 100%. Your technical team will know the difference.

- **Transceiver Cooling:** Channelized and amateur transceivers use fans or heat sinks to cool themselves which are designed with the transceiver's duty cycle specifications in mind. Even when operating at the lowest possible power, additional transceiver cooling may be necessary and with something as simple as a desk fan. Be sure to discuss these cooling issues with your technical team.

- **Use an Amplifier:** The best way to broadcast with a channelized or amateur transceiver is with a commercial duty amplifier. This will allow you then set the transmit power on the transceiver to the lowest possible setting. You can safely transmit in this way for approximately 30 to 60 minutes at a time without causing overheating damage to the transceiver.

- **Common Channel Number:** FRS/GMRS service uses channelized UHF frequencies; the CB service uses channelized HF frequencies on the 11m band. They are incompatible.

  However, a common channel number scheme is useful. FRS/GMRS and GMRS-only transceivers can send and receive on channels from 1 to 7. While these frequencies are only allowed 5W of transmit power, they have the same 25kHz bandwidth as channels, 15 to 22 with a maximum transmit power of 50W. On the other hand, channels 8 to 14 are of little value as they offer half as much bandwidth as all the other channels and only offer a maximum transmit power of 0.5W, which makes them marginal at best.

  With CB radios, 23 and 40 channel transceivers can receive and transmit on channels 1 to 23 at the same power. Therefore, a channel which is common to all of these like a channel 4 will give you a standard channel number to use for FRS/GMRS and GMRS-only, 23 and 40 channel CBs, and SSB CBs though the services are not compatible. You can simulcast then with both services on a common channel number. Everyone would know to tune to channel 4 to receive your broadcast, regardless of the service they have.

- **Microphone Overmodulation:** Aftermarket powered microphones are popular with CB operators because they improve the sound quality and range. With the unpowered microphones supplied by the factory, overmodulation can also occur when the associated potentiometer inside the transceiver is not calibrated correctly.

   Either way, an overmodulated microphone increases the transceiver's input signal above the proper level. This will result in splattering and cause annoying interference on adjacent channels or frequencies.

- **Scheduling:** As was previously discussed, the riskiest time to operate two-way radios will be during daylight hours. Therefore, your broadcasting schedule needs to protect listeners from a damaging voltage spike event while listening to your broadcast.

   For the best results, extend your "No Glow" administrative policy to your broadcast operations as well. You should not have any non-essential communications after the first glow of sunrise and before the last glow of sunset.

Remember, if you begin broadcasting before a global tribulation causes the elimination of FCC enforcement, you will face severe fines, confiscation, and possible jail time.

However, once the FCC enforcement is no longer and Emergency conditions prevail, this still does not give you a free hand to do as you please. If you lack the public's support and the support of local enforcement agencies, the local sheriff may be prompted to take action against you – unless you have alliances.

Support Your Local Sheriff

The first thing you need to do before acquiring components for a local broadcasting system is to meet with your local sheriff. You should do this at a time when only die-hard naysayers will continue to deny the onset of a global tribulation.

A sheriff will be more open-minded at this time. The emergency communications capabilities of a reliable local news network will serve a useful auxiliary role. Here is where you need to establish a trust alliance with your local sheriff to provide public service programming. This will have a direct impact on your broadcasting power. Why this particular strategy?

Let's say that you are broadcasting over CB with the FCC authorized power of 4 W. With a range of a few miles at most, there may be a few minor complaints, but relative to the other demands on the sheriff at that time, it is likely that you're not going to merit the expenditure of scarce resources.

However, hook up a 1,000 W amplifier to your channelized transceiver, and that can change very quickly. This is especially true if your system generates interference on adjacent channels due to a flawed amplifier integration design or operation.

Channelized Amplifiers

Setting up a broadcast station with amplifiers is technically challenging and expensive. However, do it right, and folks all throughout the county will be able to hear your broadcast. With the right technical team, this is doable and sustainable and here are the basic considerations:

- **Peak vs. Continuous Power:** Generally, amplifiers produce two types of transmitting power — peak and continuous. Peak power is the maximum power level the amplifier can safely sustain in short bursts. Conversely, an amplifier can transmit on a continuous basis at lower power.

  To frame this, one could think of this like a small airplane where the pilot uses more power at takeoff to get the aircraft to an altitude where the engine is then throttled back to a more sustainable cruising power.

  With channelized amplifiers, you will have a similar situation. Your amplifier will be set to a continuous transmit power level for scheduled broadcasts which is typically about 70% of peak power. However, in case of an emergency, the amplifier can be set to 100% peak power for a few minutes so that public emergency notices can be transmitted with the widest possible range.

- **Amplifier Linearity:** Linearity is the ability of an amplifier to produce signals that are accurate copies of the input at increased power levels. There are two types of amplifiers: nonlinear and linear. Linear amplifiers are preferred for two-way radios because they amplify a signal without adding distortion. The widest use for nonlinear amplifiers is for CW, Morse code transmissions where output power not fidelity, is more important.

- **Amplifier Type:** Vacuum tube vs. solid state. A well-made vacuum tube amplifier can last for 50 years, and many HAMs build these amplifiers using well-established designs. Solid state amplifiers are more popular these days with HAM operators because they cost less and consume significantly less power than vacuum tube amplifiers.

  It is possible though, that your technical team may still prefer vacuum tube amplifiers. Why? They are more likely to survive the kind of global tribulation power spike events that can destroy a solid state amplifier.

- **Antenna Tuner:** While many transceivers feature built-in antenna tuners, they are only useful when "running barefoot" (without an amplifier). For channelized broadcasting with a linear solid-state amplifier, an external and adjustable antenna tuner is preferred.

The bottom line with amplified channelized broadcasting is that an improperly configured system will cause interference to adjacent channels; so make sure your technical team tests your system thoroughly as you focus your efforts on building popular support and alliances.

Always remember; channelized transceivers are not designed for broadcast and pushing them beyond 60 minutes for a scheduled broadcast will likely cause severe damage to the equipment. Also, building trusted alliances with your local sheriff and municipalities would be essential, especially if you have your sights set on AM/FM tribulation broadcasting.

AM/FM Tribulation Broadcasting

If the sheriff or other local law enforcement agency sees your local Radio Free Earth channelized broadcasts as a valuable public service in the absence of FCC enforcement, you will likely engender support for AM/FM tribulation broadcasting as well.

As is shown in this book, your efforts to implement a community communications strategy will develop in stages until all three levels are operational. Then, the next step is channelized broadcasting, and finally, the light at the end of the tunnel – AM/FM tribulation broadcasting.

AM/FM Pros and Cons

The choice between AM or FM will depend mostly on the local terrain, but all things considered, FM is the best option. The criteria for salvaging broadcasting equipment are:

- **AM/FM Broadcast Power:** The ideal station for salvage will use a broadcast power of 5 kW (5,000 W) during the day and 2 kW (2,000 W) at night.

- **FM Stations:** Commercial FM radio frequencies range from 88 to 108 MHz. Colleges and PBS stations usually transmit in the lower frequencies and commercial stations above them. While salvaging equipment may be tricky, you can count on seeing scores of commercial stations close, as staff leave to take care of their families and as network syndication feeds fail.

- **AM Stations:** Commercial AM radio frequencies range from 530 to 1700 kHz. The downside with AM is sound quality and power as compared with commercial FM. AM is best suited for spoken-word formats such as news radio.

- **Daytime Range:** AM and FM will have approximately the same daytime range. Given a transmit power of 2 kW or more the daytime range for AM and FM stations will be about 30 miles. If you are using a "No Glow" administrative policy, you'll only broadcast for a few minutes during the day to make emergency announcements.

- **Nighttime Range:** An FM signal will still travel the same distance as it does during the day, whereas an AM signal will go approximately twice the distance or more at night, given prevailing ionospheric conditions.

The advantage of AM is the longer range at night than FM at the same power. However, FM provides a cleaner signal than AM and it is affected less by weather and lightning.

AM/FM Technical Considerations

There is a huge difference between commercial grade equipment and consumer and amateur equipment. It is so extreme that it is like comparing a golf cart to an 18 wheeler. In other words, this is not a simple thing to accomplish, though it is doable.

If you have ambitions to begin AM/FM tribulation broadcasting in the absence of FCC enforcement, you'll need to salvage equipment from an abandoned AM or FM commercial station. To you, this equipment is alien when compared to what you have learned about channelized and amateur transceivers.

To illustrate the complexities of setting up a local AM/FM tribulation broadcasting network, let's step through the technical steps required to get an electronic signal from a microphone to an antenna so that it goes out over the air.

Since there are many similarities and few differences between AM and FM broadcasting, the following offers a combined overview for both. Please note, many commercial stations use custom-built transmitters, and there can be many subtle differences following steps offer a general high-level view of the process

1. **Microphone:** The first step is to capture a voice with a microphone, which converts the sound to an electronic signal that goes to a preamplifier.

2. **Preamplifier:** Also called a preamp, the purpose of this device is to correct the audio level of the signal coming from the microphone to a standard level. Once processed, the signal is completed by the preamplifier, and it then goes to an audio processor.

3. **Audio Processor:** With a properly adjusted audio level, the audio processor's next step is to control the tone and dynamic range of the sound before it goes to the mixer.

4. **Mixer:** A broadcast station may use multiple sources, such as a studio or syndication feed to form a complete broadcast via a mixer. The mixer is the last component that processes the signal before it leaves the broadcast control room via the feedline as a single signal.

5. **Feedline:** The feedline connects the broadcast control room to the antenna site via multiple-strand fiber optic or balanced wire cables. A balanced wire feed will use two wires for mono and four wires for stereo.

   Multiple-strand fiber optic or balanced wire cables will typically have at least eight separate strands or wires. Control and engineering applications use the additional strands or wires. With copper wire, an inline powered amplifier or repeater may be necessary between the mixer and the antenna site. However, since fiber optic has a much longer range, it may not require amplifiers or repeaters. The best choice is fiber optic for two reasons: fewer components and better EMP protection.

6. **Antenna Site Entry Point:** The entry point for the feedline coming from the broadcast control room into the antenna site is the line conditioner.

7. **Line Conditioner:** The first component to process the signal in the antenna site is the conditioner or demodulator. It captures and reconditions the broadcast signal to resolve any issues that may have occurred along the feedline path to ensure that the broadcast signal meets the necessary specifications. From the line conditioner, the signals then go to a demultiplexer.

8. **Demultiplexer:** This device splits the audio, control and engineering lines from the feedline into their constituent parts for other broadcast components.

9. **Demultiplexer to Engineer:** The control and engineering signals used by a remotely located engineer are sent back along the feedline to the broadcast control room. The engineer then uses these signals to control broadcast functions and to monitor and calibrate the broadcast system's health and status.

10. **Demultiplexer to Exciter:** The actual broadcast signals from the demultiplexer then go to the exciter.

11. **Exciter:** An exciter plays a crucial role in AM/FM broadcasting. It combines the audio signals with a carrier wave to generate a complete RF signal with the desired modulation, AM or FM on the frequency assigned to the station by the FCC. It will also boost the power of the complete RF signal to the level required by the final amplifier.

12. **Final Amplifier:** There may be several amplifiers along the signal path from the exciter to the antenna. However, only one amplifier eventually feeds the antenna; this is the final amplifier.

    A final amplifier can be a single component or an interconnected array of components designed for a more efficient push-pull configuration. What is different between the final amplifier and those which precede it in the path, is the use of amplifier controls and filters.

13. **Amplifier Controls and Filters:** Other systems may be used in conjunction with a final amplifier. One is the engineering controls used to prevent damage to the amplifier from a damaged or mismatched antenna. Another could be low pass filters that remove unwanted harmonics from the signal to prevent interference in adjacent frequencies.

14. **Feedline to the Antenna:** A major difference between the feedline that connects the broadcast control station to the antenna site and the feedline that goes from the antenna site to the antenna is the amount of power. It will be huge.

    The physical distance between the final amplifier and the antenna's feed point is critical, and the shorter it is the better. Therefore, co-location between transmitter sites and the antennas they serve is important. There are significant differences between AM and FM broadcast antennas also.

15. **AM Antenna:** The use of more than one vertical AM frequency antenna may be required and insulation from the ground is required for each antenna. Because the feedline connects directly to the tower, the entire tower array becomes the complete broadcast antenna.

16. **FM Antenna:** FM antennas are much smaller than AM antennas. The physical connection between the feedline and each antenna must connect with a grounding field. Multiple antennas are mounted in a fixed pattern to the same or different tower structures to provide the desired RF pattern.

It is important to note that FM only broadcasts through an antenna; the tower is only a support structure. Therefore, the inherent design of FM broadcast antennas makes them better suited to EMP protection than AM broadcast antennas.

If your community has a hydroelectric dam, a 50 kW AM station is certainly an option, if you want to get your signal really out there. You'll get the attention of a lot more people that way, and that could have some serious downsides.

The bottom line with AM/FM broadcasting is that FM is the way to go for local broadcasting. Sound quality issues aside, there are many more FM receivers of all types than AM receivers of all types. Also, FM offers less vulnerable antenna systems and better protection from EMP, lightning, and weather.

Now that you've gone through all these steps, your eyes may be rolling back in your head as you exclaim "OMG." While it may tickle your imagination to operate an AM/FM tribulation broadcasting network, imagining how to get there can be daunting.

No worries dear reader; there is a possible shortcut.

Tribulation Broadcasting Alliances

One of the reasons why "Fake News" has become the media plague of the 21st century is that the FCC removed all of the restrictions on consolidation and syndication, thereby creating a propaganda machine windfall for the elites. Consequently, a once vibrant and competitive "Fourth Estate" has become largely a pack of agenda-driven entertainers.

One consequence of this is that our notions of a radio station's role have not kept up with the times. When we think of radio stations, we imagine studios in the town served by the station where personalities give interviews and entertainers perform. Unfortunately, this is the exception today – and not the rule.

The rule today is that one syndicated program can provide programming to hundreds of small stations across the nation. The technical implications for this have been considerable. Now, there are antenna sites where multiple competitive stations share facilities relying solely on syndication feeds.

Only a few broadcast engineers are needed to maintain these facilities while working for all of the stations. In a city of 300,000 people, the number of these broadcast engineers can be five or less.

Therefore, if you want to start a local AM/FM tribulation broadcasting station by salvaging abandoned or out-of-use equipment, the first person you need to contact is the local sheriff. A sheriff will know where these syndication feed sites are, who owns them, and the broadcast engineers who maintain them

A sheriff that supports your AM/FM tribulation broadcasting goals will personally know them or how to find them assuming they're still available.

If you can find one of them and make your case, the rest will come together on its own. That's how you get started. When you forge your alliances, be sure to put this on the table from the get-go.

If you're wondering if this is worth the effort, consider the 1958 cinema classic, South Pacific. Muriel Smith performs in the film as the singing voice of actress Anna May Wong for the song, "Happy Talk." Below is a partial lyric.

"Happy Talk" from South Pacific (1949) - Rodgers and Hammerstein (Partial Lyrics)

Happy talk, keep talking happy talk,
Talk about things you'd like to do,
You gotta have a dream if you don't have a dream,
How you gonna have a dream come true?

Coming up next: a dream worth singing…

24

Hope for the Future

Starting a local Radio Free Earth network public service broadcast is a huge step up from a self-focused "me-and-mine" sheltering strategy. This is where one survives for the sake of survival alone on the front end of a crisis. After that, things become unfocused.

Sheltering is a short-term strategy. It assumes that once you emerge from your shelter (to survive in what remains of the last civilization after a few months or years) you are smart enough to work things out as you go along.

The assumption is that a few months or perhaps a year will have passed. This assumption is based on residual and untested Cold War thinking and can only serve as a deadly blind spot. The coming global tribulation will probably last a decade or longer.

It would surprise most to learn that many of the nation's top 2% have already made their own "me-and-mine" shelter arrangements, though they will be careful to not divulge that. After all, the first rule of any "me-and-mine" sheltering strategy is to avoid unwanted guests in the future by denying everything today, especially in the case of a possible global tribulation.

Nevertheless, many of these wealthy people have constructed expensive underground deluxe shelters with up to two years of provisions, fuel, water, and so forth at the cost of one million dollars or more. How will that work out for them? In most cases, it will be a fearful existence during or after which they will die badly.

In the beginning, the back-slapping about being clever survivors will be heartfelt. This celebration will be short-lived although, after which will come nagging doubts. There will be long nights where the assumptions about how much food, fuel, and water will prove to have been much too optimistic.

What about those scraping sounds above ground? Is a local contractor who remembered the shelter's location coming back to dig you out to feed his family? Or have desperate people found an access point to your bunker, such as a vent pipe your contractor hide inside an old tree, and now they are digging you out? Or what about those awful sounds coming from the generator room after that last trembler and you're out of spare parts?

As the old saying goes, "Nothing goes according to plan" and now the over-confident "me-and-mine" shelter survivors must come to the surface much earlier than planned.

If you think the right question at this point is, "What do they find on the surface?" You're dead wrong. The right question is, "Who will find those who finally emerge on the surface?"

"Me-and-mine" sheltering planners always look for remote places which during the coming tribulation will become part of a patchwork quilt of "no man's land" areas beyond the reach of resource-limited sheriffs.

It is in these areas that the dregs of humanity will hunt "me-and-mine" sheltering survivors that are prematurely forced to the surface. The horror will begin once the dregs smell the healthy MRE-fed bodies of emerging survivors from up to a mile away. Then, they'll invite those unfortunate survivors to dinner where they will become the main course – long pig au jus.

As for those survivors who banded together in survival communities of one hundred or more and use two-way radios to help track all movement in their areas, they will be a hard and dangerous target for the dregs of humanity.

The dregs will learn that if they send scouts to reconnoiter these areas, few will return; so they'll focus on the soft targets, the "me-and-mine' sheltering survivors who have yet to learn how to survive the new above ground reality.

There will be a final reckoning in time, as good men and women hunt down the hunters. Noble warriors who have sworn to protect the innocent, the good, and the vulnerable and to cleanse the world of those who lost their humanity long before or during the tribulation. They know that the good will eventually prevail and that the stouthearted will not go down to destruction.

However, protecting others is not an easy thing when placing a predator's head in the crosshairs of a rifle scope, even after this predator has raped and murdered an innocent child. Pulling the trigger in this case may prove to be harder to do than most could imagine.

There will be noble warriors, men, and women committed to serving fellow community members. They will have learned their craft in the military or law enforcement, and they will be the ones to pull that trigger without hesitation or regret.

Therefore, the question is, what role will you perform in supporting the innocent, the good, and the vulnerable? Will you step out, step aside, or step up with hope for the future?

Stepping Up

In the coming global tribulation, there will be a continuous calling from above that urges those who believe in a new possibility for humankind. When the worst is long over, and the remaining survivors find themselves on the "backside" as I call it. It is a time when we all see clear skies and taste sweet waters once again.

Then, there will be those who enclose themselves in material self-interest, and they'll naturally view an enlightened view as foolishness and prefer a dog-eat-dog, survival of the fittest path. Provided these self-interested survivors avoid conflict this is workable though you should not pander to them or their issues. It is better to ignore or evade them.

If you desire to broadcast news and survival infotainment, then step up your efforts to focus on the needs of good people. Do that and those who support you and your belief in the backside will take on a very important role.

Survival Infotainment

Today, infotainment is a term used to describe broadcast material that combines entertainment with informative programming. During the tribulation, there will be a simpler definition.

If you cannot feed the belly, feed the soul.

How is this done? By combining programming that uplifts the soul with common sense survival tips of value to all. A few examples of survival infotainment are:

- **Music:** Catchy tunes about common survival tasks such as filtering ditch water and of course jazz with a similar useful twist.

- **Inspirational Talks:** The Great Depression was a time of great hardships for Americans. President Franklin D. Roosevelt became a voice for those suffering in 1933 during the Great Depression. He spoke to all Americans, and his warm and kind voice kindled the hopes of Americans with some 30 speeches called the "Fireside Chats."

- **Family Oriented:** Soap operas will always be popular, and these plot lines are easy to adapt to the needs of survivors. For example, a soap opera that stars a group of midwives. There will be the usual dramas, but along the way, listeners will learn useful skills such as prenatal care, difficult deliveries and so forth.

- **Announcements:** Celebrate the victories of living such as marriages, births and civic accomplishments. Good news for survivors to hear that will inspire them with hope for the future and a noble sense of continuity.

- **Humor:** In the early 1950's Americans enjoyed the Amos 'n Andy Show radio show. The gags about a Stutz Bearcat automobile were always good for a laugh. During the tribulation, survival needs can be served using the same format but with a twist. Instead of gag punch lines about an old car, the comedians can center their humor on common survival tasks such as digging a well or building a raised bed garden.

What format works best? Where should you begin? You begin by listening to your audience for what kind of survival infotainment will work best for them.

No matter what you do eventually, it must embrace the following three survival precepts:

- **Self-Sufficiency:** Independence is the foundation of spiritual freedom, and without self-sufficiency, there is little chance of independence.

- **Hope for the Future:** Lamenting the end of this civilization will only sap one's will to survive. Believing in a more enlightened future will give survivors something worth living for and a reason to get up in the morning to face another day of hardship.

- **Knowing You Are Not Alone:** Joining with like-minded others through positive alliances will create consensus. Those who seek a more enlightened future must actively do so.

We've covered a lot of ground in this book, and you may wonder where I first got the idea for it? This question makes me smile. Because, even though I've spent over two years researching and authoring this book, the inspiration for it predates all of that.

Inspiration Half a World Away

In 1992, I started a tour business specializing in independent travel and homestays in the former states of the Soviet Union. It was a time when Russians and Americans were equally interested in getting to know each after decades of living in fear of each other's ballistic missiles, and homestays were the perfect way to look beyond the decades of fear and prejudice that had separated our respective nations.

The 1990's were good times for the relations between our two nations, and back then, our state department was bullish on the new Russian Federation. In fact, they went to great lengths to encourage Americans like me to do business there, but frankly, my real motivation was more of a mix of adventure and curiosity.

My first trip to Russia was in the summer of 1992 on Lufthansa, and that is a long trip when you're flying in "spam-in-can" class. Because of my specialty, Aeroflot gave me the same discounted rates reserved for travel agents; so I began flying the polar route from San Francisco to Moscow to conduct site visits and to work with my operators. The Russian passenger jets were also much more comfortable to fly in than Western-made jets.

To prepare for the summer travel season, I would fly over in December and return in January and did this from 1992 to 1998. The flights from San Francisco to Moscow were at night and during the day on the return legs.

My first trip over was in an Ilyushin Il-62 in December 1992 and back again in January 1993. It was one of the most remarkable days of my life, and it was the first time I ever saw blue ice which is very old ice.

Flying above the top of the world, I was mesmerized by this sight of pristine and unbroken ice passing beneath the jet airliner. White, blue and virtually unbroken it was a stunning sight.

Then, over the years, as I continued flying the same winter polar route at the same time with Aeroflot, something sad happened.

I watched in dismay as the beautiful solid ice cap I first saw in January 1993 began deteriorating consistently year-after-year until my final return flight in 1998. The ice cap had begun to look like the shattered windshield of a junked car in a wrecking yard.

It was during that last return polar flight that I concluded that our world was changing for the worse. Instead of allowing special interests to tell me what to believe or to mock me for not embracing their lies, I decided to trust my own eyes and have done so ever since. If current Earth changes tell us anything, I made the right call, though I suffered a lot of abuse for it.

Looking back on those days, I learned that what you get out of adventure is not always what you expected, and this was not the only unexpected result.

Voice of America

My bachelor's degree was in communications. During the 1980's, I worked as contract science features producer for CNN and as a technical commentator for the Texas Cable Network. Consequently, I've always held an interest in the effectiveness of media from the audience's point of view.

During the Cold War, there was a dogged fear of a nuclear confrontation which sadly persists to this day. Consequently, a dominant part of the Cold War was propaganda.

The Soviets had Pravda, and the West competed with that message on the Voice of America (VOA), Radio Liberty (RL), and the BBC. All three, VOA, RL, and BBC were government funded networks.

This Cold War battle of the airwaves was on my mind when I booked homestays with professional families. Therefore, I always enjoyed speaking with my hosts about those days, and while each family had their own unique experiences, there were common threads that coursed through every conversation.

First was, everything was seen as propaganda to the Russians. Whether the news was broadcast by their Soviet state or by the West made no difference. That is except for one notable exception. As one pithy Muscovite (a man born in Moscow, Russia) told me, "Voice of America was always useful propaganda."

Because of its integrity, the Voice of America (VOA) was the hands-down favorite among Soviet citizens. Based on my conversations, what I determined over time was that VOA had roughly 70% of the listener audience. Radio Liberty (RL) and the BBC came in a distant second and third place, and they had to compete for the remaining 30% of the audience.

Why VOA Was Popular

The Voice of America (VOA) was founded on February 1, 1942. It began to broadcast to Russia in AM, FM, and Shortwave frequencies in the Russian language on February 17, 1947. Russians heard on that day and for the very first time, "Hello! This is New York calling. You are listening to the first radio broadcast of Voice of the United States of America."

With an audience of 30 million listeners each week, VOA was clearly the dominant voice of the West in Russia during the Cold War and for three principal reasons: trustworthiness, viewpoints, and entertainment.

Regarding trustworthiness, the BBC's and Radio Liberty's political agendas and tactics were visible to savvy Russians. However, the VOA's news reporting showed a consistent effort to gain and keep the listener's trust.

The Russians appreciated this about VOA, but what made them love VOA broadcasts were all the interesting things to listen to besides the news. They could listen to people discussing different political views and hear readings of Alexander Solzhenitsyn's famous book, *The Gulag Archipelago,* for example.

Then, there were the music programs and here is where VOA shined. Every Russian I spoke with about VOA always told me how much they loved the music, and why their broadcasts of Jazz, swing, and early bebop were so deeply appreciated.

These moments in the conversations were always precious to me because I could see folks remembering those days of risking imprisonment to hear jazz on the VOA. Their faces would always soften with a warm smile.

The lesson here for broadcasting during a global tribulation is that if you want to build an audience, agenda-driven newscasts and boring entertainment will not take you very far.

Rather, if you want to build a relationship with your audience, find their passion, what they love, and the human interest angles that attract them, and program for it.

Not only was this a significant lesson about the success of VOA, but there was another lesson, which would be immensely important to broadcasting during a global tribulation.

The Invasive Power of RF

Did the Soviet government actively attempt to jam the VOA, RL, and BBC? Absolutely, and they spent a fortune doing so. Their efforts to jam overseas broadcasts aimed at their country were not universally effective because their efforts depended on how many radio jammers they could put into a given area. The result was that while heavy jamming was possible in the big cities, in the countryside it was significantly less effective because there was a lower density of radio jammers. This resulted in the rural Russians finding it much easier to receive the broadcasts.

In cities those who could afford the cost, while avoiding the scrutiny of the KGB and the police, would buy black market shortwave radios which proved to be less susceptible to jamming. Or, they could go through a lot of bureaucracy to get one and lose their anonymity in the process - (much like we do today with Facebook). Same old problem – Uncle Ivan sees you, though Russians invariably found ways to tune in, and they did – in the millions.

The point here is that the elites are not all-powerful. They can impede the ability of communities to broadcast hope for the future, but they cannot block it. If what you have is worth listening to for the people they want to keep in the dark, those people will find a way out of the dark to you and your message.

Never forget that the real power of the Radio Frequency (RF) spectrum is that it is invasive.

One of the things that also makes me sad is the current trend in amateur radio towards dependence upon digital infrastructure. This new generation of technology is sweeping away the longest standing and the single greatest virtue of amateur radio: its independence from infrastructure. Therefore, if you want the invasive power of RF to work for you and your network, go and stay analog.

We now arrive at the final question. Given all the effort and resources needed to broadcast news and survival infotainment, is this worth it?

Let's be honest. Most will say that such an effort is a "bridge too far" as the saying goes, and therefore, a waste of time and resources. So dear reader, are you most people? Or, are you willing to march to the beat of a different drummer?

Are you ready to join the courageous freedom lovers of each generation who gave their all and sometimes their lives for the noble purpose of freedom?

If at the moment you do not have the answer to this question, then give it to God.

If you are sincere, synchronicity will cause power opportunities to appear, and when they do, remember the wise words of Louis Pasteur, "Chance favors the prepared mind." Seize them, and more will come. Believe it.

Appendices

Appendix A – Radio Codes

International Morse Code

- The length of a dot is one unit.
- A dash is three units.
- The space between parts of the same letter is one unit.
- The space between letters is three units.
- The space between words is seven units.

International Morse Code Letters			
A	● ▬	N	▬ ●
B	▬ ● ● ●	O	▬ ▬ ▬
C	▬ ● ▬ ●	P	● ▬ ▬ ●
D	▬ ● ●	Q	▬ ▬ ● ▬
E	●	R	● ▬ ●
F	● ● ▬ ●	S	● ● ●
G	▬ ▬ ●	T	▬
H	● ● ● ●	U	● ● ▬
I	● ●	V	● ● ● ▬
J	● ▬ ▬ ▬	W	● ▬ ▬
K	▬ ● ▬	X	▬ ● ● ▬

International Morse Code Letters

L	• ▬ • •	Y	▬ • ▬ ▬
M	▬ ▬	Z	▬ ▬ • •

International Morse Code Numbers

1	• ▬ ▬ ▬ ▬
2	• • ▬ ▬ ▬
3	• • • ▬ ▬
4	• • • • ▬
5	• • • • •
6	▬ • • • •
7	▬ ▬ • • •
8	▬ ▬ ▬ • •
9	▬ ▬ ▬ ▬ •
0	▬ ▬ ▬ ▬ ▬

International Morse Code Special Letters

Ä	• ▬ • ▬
Á	• ▬ ▬ • ▬
Å	• ▬ ▬ • ▬
Ch	▬ ▬ ▬ ▬

International Morse Code Special Letters	
É	● ● ▬ ● ●
Ñ	▬ ▬ ● ▬ ▬
Ö	▬ ▬ ▬ ●
Ü	● ● ▬ ▬

International Morse Code Punctuation Marks		
Character	**Description**	**Morse Code**
.	Full-stop (period)	● ▬ ● ▬ ● ▬
,	Comma	▬ ▬ ● ● ▬ ▬
?	Question mark	● ● ▬ ▬ ● ●
'	Apostrophe	● ▬ ▬ ▬ ▬ ●
!	Exclamation	▬ ● ▬ ● ▬ ▬
/	Slash	▬ ● ● ▬ ●
(	Open Parentheses	▬ ● ▬ ▬ ●
)	Close Parentheses	▬ ● ▬ ▬ ● ▬
&	Ampersand	● ▬ ● ● ●
:	Colon	▬ ▬ ▬ ● ● ●
;	Semicolon	▬ ● ▬ ● ▬ ●
=	Equals sign	▬ ● ● ● ▬
-	Hyphen	▬ ● ● ● ● ▬
+	Plus Sign	● ▬ ● ▬ ●

International Morse Code Punctuation Marks		
Character	**Description**	**Morse Code**
–	Minus Sign	● ● ▬ ▬ ● ▬
"	Quotation Marks	● ▬ ● ● ▬ ●
$	Dollar Sign	● ● ● ▬ ● ● ▬
@	At Sign	● ▬ ▬ ● ▬ ●

Morse Code Abbreviations

This partial listings of Morse code abbreviations shown below are used to shorten CW transmission times.

International Morse Code Abbreviations			
BTU	Back to You	**K**	Over
CQ	Attention All Stations	**NX**	Noise; noisy
CS	Callsign	**OB**	Old Boy
DE	This is...	**PLS**	Please
DSW	Goodbye	**R**	Roger
DX	Long Distance Contact	**TNX**	Thanks
EMRG	Emergency	**UR**	You're
GE	Good evening	**73**	Best Regards
GL	Good luck	**77**	Long Live Morse Code
GM	Good morning	**88**	Love and Kisses
GN	Good night	**99**	Get Lost!
II	I say again		

Prosign and Q Codes

This partial listings of Prosign informal Q codes are used to shorten CW transmission times. Note: Q codes are sometimes used with voice transmissions.

Prosign Codes	
Code	**Description**
AA	New line
AR	End of message
AS	Wait
BK	Break
BT	New paragraph
CL	Going off the air ("clear")
CT	Start copying
KN	Invite a specific station to transmit
SK	End of transmission
SN	Understood
SOS	Distress message

Q Codes (Partial Listing)	
Code	**Description**
QRA	Name
QRB	Distance
QRG	Frequency
QRK	Intelligibility
QRL	Busy
QRM	Interference
QRN	Noise
QRO	High power
QRP	Low power
QRQ	High speed CW
QRRR	Land SOS
QRS	Low speed CW
QRSS	Very low speed CW
QRT	Shut down the station
QST	Attention all amateurs
QRV	Ready
QRX	Stand by
QRZ	Who is calling
QSB	Fading
QSD	Defective keying

Q Codes (Partial Listing)	
QSK	Break in
QSL	Confirmation
QSO	Radio contact
QSY	Change frequency
QTC	Message
QTH	Location
QTR	Time

International NATO Phonetic Alphabet

Standard international phonetic alphabet used with military and aviation voice transmissions.

NATO Phonetic Alphabet			
A	Alfa	N	November
B	Bravo	O	Oscar
C	Charlie	P	Papa
D	Delta	Q	Quebec
E	Echo	R	Romeo
F	Foxtrot	S	Sierra
G	Golf	T	Tango
H	Hotel	U	Uniform
I	India	V	Victor
J	Juliett	W	Whiskey
K	Kilo	X	X-ray
L	Lima	Y	Yankee
M	Mike	Z	Zulu

RST Codes

There are three different RST codes (readability, strength and tone) used by amateur radio operators to describe the quality of a radio signal that is being received. Note:

- For phone mode, readability and strength are used.
- For CW and digital modes you use all three, readability, strength and tone.

RST codes are used to describe quality of a radio signal being received.

RST Values for Readability	
Value	**Code Description**
1	Unreadable
2	Barely readable, occasional words distinguishable
3	Readable with considerable difficulty
4	Readable with practically no difficulty
5	Perfectly readable

RST Values for Strength	
Value	**Code Description**
1	Faint signal, barely perceptible
2	Very weak
3	Weak
4	Fair
5	Fairly good
6	Good
7	Moderately strong
8	Strong
9	Very strong signals

RST Values for Tone (CW and Digital Only)	
Value	**Code Description**
1	Sixty cycle a.c or less, very rough and broad
2	Very rough a.c., very harsh and broad
3	Rough a.c. tone, rectified but not filtered
4	Rough note, some trace of filtering
5	Filtered rectified a.c. but strongly ripple-modulated
6	Filtered tone, definite trace of ripple modulation
7	Near pure tone, trace of ripple modulation
8	Near perfect tone, slight trace of modulation

RST Values for Tone (CW and Digital Only)	
9	Perfect tone, no trace of ripple or modulation of any kind

10 Codes

Ten-codes are brevity codes. Officially known as ten signals, they represent common voice communication phrases, used in law enforcement and Citizens Band (CB) radio transmissions. Note, the values of the 10-codes used by law enforcement can are typically used for internal radio transmissions and can vary between jurisdictions.

10 Codes for CB and Law Enforcement	
Value	**Code Description**
10-1	Received poorly
10-2	Receiving well
10-3	Stop transmitting
10-4	OK, message received
10-5	Relay message
10-6	Busy, stand by
10-7	Out of service, leaving air
10-8	In service, subject to call
10-9	Repeat message
10-10	Transmission completed, standing by
10-11	Talking too rapidly
10-12	Visitors present
10-13	Advise Weather/ Road conditions
10-16	Make pickup at
10-17	Urgent business
10-18	Anything for us?

10 Codes for CB and Law Enforcement	
10-19	Nothing for you, return to base
10-20	My location is
10-21	Call by telephone
10-22	Report in person to
10-23	Stand by
10-24	Completed last assignment
10-25	Can you contact
10-26	Disregard last information
10-27	I am moving to channel
10-28	Identify your station
10-29	Time is up for contact
10-30	Does not conform to FCC rules
10-31	Crime in Progress
10-32	I will give you a radio check
10-33	EMERGENCY TRAFFIC
10-34	Trouble at this station
10-35	Confidential information
10-36	Correct time is
10-37	Wrecker needed at
10-38	Ambulance needed at
10-39	Your message is delivered

10 Codes for CB and Law Enforcement	
10-40	Silent Run – No Light, Siren
10-41	Please turn to channel
10-42	Traffic accident at
10-43	Traffic tie up at
10-44	I have a message for you
10-45	All units within range please report
10-46	Assist Motorist
10-47	Emergency Road Repair at
10-48	Traffic Standard Repair at
10-49	Traffic Light Out at
10-50	Break channel
10-60	What is next message number
10-62	Unable to copy, use phone
10-63	Net directed to
10-64	Net clear
10-65	Awaiting your next message
10-67	All units comply
10-70	Fire at
10-71	OK to transmit in sequence
10-77	Negative contact
10-81	Reserve hotel room for

10 Codes for CB and Law Enforcement	
10-82	Reserve hotel room for officer
10-84	My telephone number is
10-85	My address is
10-91	Talk closer to microphone
10-93	Check my frequency on this channel
10-94	Please give me a long count
10-99	Mission completed, all units secure
10-100	Need to go to Bathroom
10-200	Police needed at

Appendix B – Titanic Radio Log for April 1912

Titanic Wireless Antenna

Titanic Radio Log			
Time	**Sender**	**Message**	**Notes**
		14 April 1912	
2050	**Californian**		*Cyril Evans, the radio operator of the Californian, a British Leyland Line steamship, is transmitting iceberg warnings to all stations.*
			Titanic signals jammed by Californian due to proximity with Titanic.
2055	**Titanic**	TITANIC KEEP OUT SHUT UP I'M WORKING CAPE RACE	*Sent to Californian*
2056			*Cyril Evans, the only operator on the Californian, shuts down the radio room for the night and goes to bed.*

Time	Sender	Message	Notes
2340			*Titanic Strikes an iceberg on its maiden voyage.* *Immediately following the collision, Captain Smith, believed the Titanic would remain afloat for hours and that it might not sink.*
		15 April 1912	
0015	**Titanic**	TITANIC TO CAPE RACE: TO HARRISON SANDFORD, NEW YORK HELLO BOY. DINING WITH YOU IN SPIRIT TONIGHT. HEART WITH YOU ALWAYS. BE LOVE, GIRL. TITANIC TO CAPE RACE: GEORGE SIMUND, NEW YORK WEATHER DELIGHTFUL. FEELING FINE. HOPE ALL	*Cape Race is a wireless coast station on the southern tip of Newfoundland's Avalon Peninsula.* *This coast station played a pivotal role in the rescue of Titanic survivors.*
	Titanic	CQD THIS IS TITANIC CQD THIS IS TITANIC CQD THIS IS TITANIC CQD THIS IS TITANIC CQD THIS IS TITANIC CQD THIS IS TITANIC POSITION 41.44N 50.24W	*Titanic's call sign was MGY. and CQD stood for Attention All Stations: Distress. The actual code transmitted was CQD DE MGY, for CQD THIS IS TITANIC.*
	Frankfurt	FRANKFURT TO TITANIC: WHAT IS THE MATTER?	

Titanic Radio Log			
Time	**Sender**	**Message**	**Notes**
	Titanic	CQD THIS IS TITANIC POSITION 41.44N 50.24	
	Frankfurt	OK. STAND BY	
	Mount Temple	MOUNT TEMPLE TO TITANIC WHAT IS THE MATTER?	
	Titanic	TITANIC TO MOUNT TEMPLE: CANNOT READ YOU OLD MAN BUT HERE MY POSITION 41.46N 50.24W COME AT ONCE. HAVE STRUCK A BERG.	
	Mount Temple	RECEIVED WILL TELL CAPTAIN	
	Titanic	CQD THIS IS TITANIC CQD CQD	
	Mount Temple	ATTENTION ALL STATIONS: THIS IS MOUNT TEMPLE TITANIC SENDING CQD. SAYS REQUIRES ASSISTANCE. GIVES POSITION. CANNOT HEAR ME. ADVISE MY CAPTAIN HIS POSITION AT 41.46N 50.24W	
	Cape Race	ATTENTION ALL STATIONS: THIS IS CAPE RACE TITANIC GIVING POSITION ON CQD POSITION 41.44N 50.24W	
0018	**Ypiranga**	ATTENTION ALL STATIONS: THIS IS YPIRANGA. TITANIC GIVES CQD HERE: 41.44N 50.24W REQUIRE ASSISTANCE	*Message Sent Ten Times*
0025	**Carpathia**	CARPATHIA TO TITANIC: DO YOU KNOW THAT CAPE COD IS SENDING A BATCH OF MESSAGES FOR YOU?	
	Titanic	TITANIC TO CARPATHIA: COME AT ONCE. WE HAVE STRUCK A BERG. IT'S A CQD OLD MAN. POSITION 41.46N 50.14W	

		Titanic Radio Log	
Time	Sender	Message	Notes
	Carpathia	CARPATHIA TO TITANIC: SHALL I TELL MY CAPTAIN? DO YOU REQUIRE ASSISTANCE?	
	Titanic	YES. COME QUICK	
	Titanic	CQD REQUIRE ASSISTANCE CORRECTED POSITION 41.46N 50.14W STRUCK ICEBERG. CQD	
	Cape Race	ATTENTION ALL SHIPS: THIS IS CAPE RACE TITANIC GIVES CORRECTED POSITON 41.46N 50.14W CALLING HIM. NO ANSWER.	
	Burma	BURMA TO TITANIC: WHAT IS YOUR POSITION?	
	Titanic	POSITION 41.46N 50.14W REQUIRE IMMEDIATE ASSISTANCE WE HAVE COLLISION WITH AN ICEBERG SINKING. CAN HEAR NOTHING FOR NOISE OF STEAM.	
0026	Titanic	TITANIC TO YPIRANGA HERE CORRECTED POSITION 41.46N 50.14W REQUIRE IMMEDIATE ASSISTANCE WE HAVE COLLISION WITH ICEBERG SINKING. CAN HEAR NOTHING FOR NOISE OF STEAM.	*Message Sent 15-20 Times*
0027	Titanic	TITANIC CQD I REQUIRE ASSISTANCE IMMEDIATELY. STRUCK BY ICEBERG IN 41.46N 50.14W	
0030	Caronia	CARONIA TO ALL SHIPS: CARONIA TO BALTIC: CQD TITANIC STRUCK ICEBERG REQUIRE IMMEDIATE ASSISTANCE.	
	Mount Temple	MOUNT TEMPLE TO TITANIC: OUT CAPTAIN REVERSES SHIP. WE ARE ABOUT 50 MILES OFF.	
0034	Frankfurt	FRANKFURT TO TITANIC POSITION 39.47N 50.10W	

		Titanic Radio Log	
Time	Sender	Message	Notes
	Titanic	TITANIC TO FRANKFURT: ARE YOU COMING TO OUR ASSISTANCE?	
	Frankfurt	FRANKFURT TO TITANIC WHAT IS THE MATTER WITH YOU?	
	Titanic	TITANIC TO FRANKFURT: WE HAVE STRUCK AN ICEBERG AND SINKING. PLEASE TELL CAPTAIN TO COME.	
	Frankfurt	FRANKFURT TO TITANIC OK WILL TELL THE BRIDGE RIGHT AWAY.	
	Titanic	OK, YES QUICK.	
	Titanic	TITANIC TO CARPATHIA: WE REQUIRE IMMEDIATE ASSISTANCE.	
	Carpathia	CARPATHIA TO TITANIC OLD MAN WE ARE 58 MILES OFF.	
	Titanic	ALRIGHT OLD MAN	
	Olympic	OLYMPIC TO TITANIC: SENDING SERVICE MESSAGE OLYMPIC TO TITANIC	
	Carpathia	CARPATHIA TO TITANIC: DON'T YOU HEAR OLYMPIC CALLING YOU?	
	Titanic	TITANIC TO CARPATHIA NO OLD MAN I CAN'T READ HIM FOR RUSH OF AIR AND NOISE OF STEAM.	
0040			*Lifeboat #7 is launched from Titanic*
	Carpathia	CARPATHIA TO TITANIC: PUTTING ABOUT AND HEADING FOR YOU. EXPECT TO ARRIVE IN FOUR HOURS.	
	Titanic	TITANIC TO CARPATHIA RECEIVED THANKS OLD MAN.	
0043			*Lifeboats #3 and #5 are launched from Titanic.*

		Titanic Radio Log	
Time	**Sender**	**Message**	**Notes**
0045			*Crew of the SS Californian see signal rockets from the Titanic and waked their captain. Captain ignored the rockets and returned to bed.*
	Titanic	TITANIC TO OLYMPIC: SOS	
	Titanic	CQD SOS THIS IS TITANIC WE HAVE STRUCK ICEBERG. SINKING FAST. COME TO OUR ASSISTANCE.	
	Titanic	TITANIC TO CALIFORNIAN: CQD CQD STRUCK ICEBERG. SINKING. REQUIRE IMMEDIATE ASSISTANCE.	*The SS Californian, was the closest ship in the area to Titanic. It's radio room was closed for the night and the Captain ignored the Titanic distress rockets.*
	Cape Race	CAPE RACE TO CALIFORNIAN SOS FROM TITANIC CQD IN 41.46N 50.14W WANTS IMMEDIATE ASSISTANCE	
0050	**Titanic**	CQD CQD THIS IS TITANIC I REQUIRE IMMEDIATE ASSISTANCE. POSITION 41.46N 50.14W	
	Caronia	CARONIA TO BALTIC: SOS CQD TITANIC IN 41.46N 50.14W WANTS IMMEDIATE ASSISTANCE.	
	Baltic	BALTIC TO CARONIA: RECEIVED	
0053	**Mount Temple**	MOUNT TEMPLE TO ALL SHIPS: TITANIC CALLING SOS	
	Titanic	CQD CQD SOS SOS SOS CQD CQD THIS IS TITANIC 41.46N 50.14W	

Titanic Radio Log			
Time	**Sender**	**Message**	**Notes**
	Olympic	OLYMPIC TO TITANIC: WHO HAS STRUCK AN ICEBERG?	
	Titanic	WE HAVE STRUCK AN ICEBERG. POSITION 41.46N 50.14W TELL CAPTAIN.	
	Olympic	RECEIVED OK	
0100			*Lifeboat #8 is launched from Titanic.*
0102	Titanic	CQD CQD SOS SOS	
0103	Titanic	TITANIC TO ASIAN: WANT IMMEDIATE ASSISTANCE. POSITION 41.46N 50.14W	
	Asian	ASIAN TO TITANIC: RECEIVED	
	Virginian	VIRGINIAN TO TITANIC	
	Cape Race	CAPE RACE TO VIRGINIAN REPORT TO YOUR CAPTAIN TITANIC HAS STRUCK ICEBERG AND REQUIRES IMMEDIATE ASSISTANCE	
	Virginian	VIRGINIAN TO CAPE RACE: RECEIVED	
			Multiple simultaneous transmissions.
			Signals jammed.
0105			*Lifeboat #1 is launched from Titanic*
	Carpathia	CARPATHIA TO TITANIC: WE ARE COMING YOUR WAY COMING AT FULL SPEED DOING 15 KNOTS	

Titanic Radio Log			
Time	**Sender**	**Message**	**Notes**
	Titanic	RECEIVED	
			Signals jammed.
0110			*Lifeboat #6 is launched from Titanic*
	Olympic	OLYMPIC TO TITANIC	
	Titanic	TITANIC TO OLYMPIC WE ARE IN COLLISION WITH A BERG SINKING HEAD DOWN	
			Signals jammed.
			Signals jammed.
			Signals jammed.
	Olympic	OLYMPIC TO ALL STATIONS	
			Signals jammed.
	Olympic	STOP TALKING	
			Signals jammed.
	Olympic	STOP TALKING STOP TRANSMITTING JAMMING ALL STATIONS STOP TALKING	
	Titanic	TITANIC TO OLYMPIC: CAPTAIN SAYS GET YOUR BOATS READY WHAT IS YOUR POSITION?	
0015	**Baltic**	BALTIC TO CARONIA: PLEASE TELL TITANIC WE ARE MAKING TOWARDS HER. WE ARE 243 MILES EAST OF TITANIC.	

Titanic Radio Log			
Time	**Sender**	**Message**	**Notes**
0020	Virginian	VIRGINIAN TO CAPE RACE: WE ARE GOING TO TITANIC'S ASSISTANCE. WE ARE 170 MILES NORTH OF TITANIC.	
	Cape Race	CAPE RACE TO TITANIC: VIRGINIAN IS GOING TO YOUR ASSISTANCE. THEIR POSITION 170 MIULES NORTH OF TITANIC.	
	Titanic	RECEIVED	
			Lifeboat #16 is launched from Titanic.
0125	Caronia	CARONIA TO TITANIC: BALTIC COMING TO YOUR ASSISTANCE.	
	Titanic	TITANIC TO BALTIC: CAPTAIN SMITH SAYS GET ALL YOUR BOATS READY. SINKING.	
	Caronia	THIS IS CARONIA TITANIC WE ARE MAKING FOR YOU. KEEP IN TOUCH WITH US.	
	Titanic	RECEIVED	
0127	Olympic	OLYMPIC TO TITANIC POSITION 40.52N 61.18W ARE YOU STEERING SOUTHERLY TO MEET US?	
	Titanic	TITANIC TO OLYMPIC WE ARE PUTTING THE WOMEN OFF IN THE BOATS.	
	Olympic	RECEIVED	
0130			*Lifeboats #9 and #12 are launched from Titanic.*

		Titanic Radio Log	
Time	**Sender**	**Message**	**Notes**
	Titanic	WE ARE PUTTING THE WOMEN OFF IN THE BOATS. WE ARE PUTTING PASSENGERS OFF IN SMALL BOATS. WOMEN AND CHILDREN IN BOATS. CANNOT LAST MUCH LONGER. LOSING POWER.	
0135	Olympic	OLYMPIC TO TITANIC WHAT WEATHER HAVE YOU HAD?	
	Titanic	CLEAR AND CALM	
			Lifeboat #11 is launched from Titanic.
	Titanic	THIS IS TITANIC ENGINE ROOM GETTING FLOODED ENGINE ROOM GETTING FLOODED	
	Frankfurt	FRANKFURT TO TITANIC ARE THERE ANY BOATS AROUND YOU ALREADY?	
	Titanic	CQD SOS CQD SOS CQD SOS ENGINE ROOM FLOODED ENGINE ROOM FLOODED CQD SOS CQD SOS	
0137	Baltic	BALTIC TO TITANIC BALTIC COMING 200 MILES EAST WE ARE RUSHING TO YOU	
0140	Olympic	OLYMPIC TO TITANIC AM LIGHTING UP ALL POSSIBLE BOILERS AS FAST AS WE CAN.	
			Lifeboat #13 is launched from Titanic.

Time	Sender	Message	Notes
	Cape Race	CAPE RACE TO VIRGINIAN: PLEASE TELL YOUR CAPTAIN THIS: OLYMPIC IS MAKING ALL SPEED FOR TITANIC BUT OLYMPIC IS 500 MILES AWAY FROM HER. VIRGINIAN YOU ARE MUCH CLOSER TO TITANIC TITANIC IS ALREADY PUTTING WOMEN OFF IN THE BOATS AND HE SAYS THE WEATHER THERE IS CALM AND CLEAR. THE OTHERS MUST BE A LONG WAY FROM TITANIC.	
	Virginian	VIRGINIAN TO CAPE RACE OK. PUTTING ON SPEED.	
	Titanic	CQD CQD SOS SOS THIS IS TITANIC	
	Carpathia	CARPATHIA TO TITANIC: BALTIC COMING TO YOUR ASSISTANCE.	
0145			*Lifeboat #2 is launched from Titanic.*
	Titanic	TITANIC TO CARPATHIA COME AS QUICKLY AS POSSIBLE OLD MAN. THE ENGINE ROOM IS FILLING UP TO THE BOILERS.	
	Carpathia	CARPATHIA TO TITANIC: RECEIVED	
0147	Frankfurt	FRANKFURT TO TITANIC ARE THERE ANY BOATS AROUND YOU ALREADY	
0148	Titanic	CQD CQD CQD SOS SOS SOS THIS IS TITANIC	
	Carpathia	CARPATHIA TO TITANIC:	
	Burma	BURMA TO TITANIC:	

		Titanic Radio Log	
Time	**Sender**	**Message**	**Notes**
	Titanic	TITANIC TO CARPATHIA COME QUICK. SHE'S TAKING ON WATER. IT'S FULL UP TO THE BOILERS.	
	Carpathia	CARPATHIA TO TITANIC: ALL OUR BOATS ARE READY. WE ARE COMING AS HARD AS WE CAN OLD MAN. DOUBLE WATCH ON ENGINE ROOM. HAVE YOUR LIFEBOATS READY WHEN WE ARRIVE.	
	Titanic	COME AS QUICKLY AS POSSIBLE OLD MAN. CQD SOS SOS SOS CQD	
0150			*Lifeboats #4 and #10 are launched from Titanic.*
	Asian	ASIAN TO TITANIC:	
	Frankfurt	FRANKFURT TO TITANIC: WE ARE 100 MILES OFF. WHAT IS THE MATTER WITH YOU?	
	Titanic	FOOL. YOU FOOL. STAND BY. STAND BY. STAND BY AND KEEP OUT. KEEP OUT.	
	Baltic	BALTIC TO CARONIA: WE ARE HEADING FOR TITANIC BUT I CAN'T AGREE TO SIGNALS.	
	Caronia	CARONIA TO BALTIC: TITANIC GIVES CQD AND SOS HER ENGINE ROOM IS FILLING UP TO THE BOILERS.	
	Asian	ASIAN TO CAPE RACE: HAVE CALLED TITANIC BUT NO REPLY. HE CANNOT HEAR ME.	
0155	**Baltic**	BALTIC TO CAPE RACE TITANIC SIGNAL VERY WEAK. DO YOU HAVE NEWS OF TITANIC?	
	Cape Race	CAPE RACE TO BALTIC: WE HAVE NOT HEARD FROM TITANIC FOR ABOUT HALF AN HOUS. HIS POWER MAY BE GONE.	

Titanic Radio Log			
Time	**Sender**	**Message**	**Notes**
0200			*Collapsible Boat C Is launched from Titanic*
	Titanic	CQD CQD THIS IS TITANIC	
0205			*Collapsible Boat D Is launched from Titanic*
0210	**Titanic**	V V CQD TITANIC WE ARE SINKING FAST. PASSENGERS ARE BEING PUT INTO BOATS. TITANIC	
	Virginian	VIRGINIAN TO TITANIC: CANNOT READ YOUR SIGNAL YOU NEED TO TRY YOUR EMERGENCY SET.	
	Titanic	SOS SOS CQD CQD TITANIC	
0217	**Titanic**	CQD THIS IS TITANIC CQD THIS IS	
			End of Transmissions
0220		*Titanic breaks apart and sinks with over 1,000 people still on board.* *Of the 2,240 passengers and crew on board, more than 1,500 lost their lives in the disaster.*	

Appendix C – IP and JIS Specifications

Two way radios will be used in a diverse range of environments and it is important to now who well they will hold up to exposure to water and dust, because there are different levels of water-proofing and dust proofing.

When evaluating radios for your community communications strategy, two reliable specifications are available: JIS and IP.

Japan Industrial Standards (JIS)

The JIS scale is commonly used with consumer two way radios and measures water resistance with levels ranging from 0-8.

JIS Codes	
Code	**Description**
JIS-0	No special protection
JIS-1	Vertically dripping water shall have no harmful effect (Drip resistant 1)
JIS-2	Dripping water at an angle up to 15 degrees from vertical shall have no harmful effect (Drip resistant 2)
JIS-3	Falling rain at an angle up to 60 degrees from vertical shall have no harmful effect (Rain resistant)
JIS-4	Splashing water from any direction shall have no harmful effect (Splash resistant)
JIS-5	Direct jetting water from any direction shall have no harmful effect (Jet resistant)Direct jetting water from any direction shall not enter the enclosure

JIS Codes	
Code	**Description**
	(Water tight)
JIS-6	Direct jetting water from any direction shall not enter the enclosure (Water tight)
JIS-7	Water shall not enter the enclosure when it is immersed in water under defined conditions (Immersion resistant)
JIS-8	The equipment is usable for continuous submersion in water under specified pressure (Submersible)

IP Code (International Protection Rating)

The IP code is commonly used with business radios to rate their resistance to both liquids and solids. An IP rating has two digits. The first digit rates resistance to solids to include fine particles such as dust and larger objects. The second digit refers to resistance to liquids.

For example, an IP rating of "IP54" assigns a rating of 5 for resistance to solids and a rating of 4 for resistance to liquids.

IP Code Solids Resistance Levels (First Digit)	
Code	**Description**
0	No protection against contact and ingress of objects
1	Any large surface of the body, such as the back of a hand, but no protection against deliberate contact with a body part (>50 mm)
2	Fingers or similar objects (>12.5 mm)
3	Tools, thick wires, etc. (>2.5 mm)
4	Most wires, screws, etc. (>1 mm)
5	Ingress of dust is not entirely prevented, but it must not enter in sufficient quantity to interfere with the satisfactory operation of the equipment; complete protection against contact
6	No ingress of dust; complete protection against contact

| IP Code Liquids Resistance Levels (Second Digit) ||
Code	Description
0	Not protected
1	Dripping water (vertically falling drops) shall have no harmful effect.
2	Vertically dripping water shall have no harmful effect when the enclosure is tilted at an angle up to 15 degrees from its normal position.
3	Water falling as a spray at any angle up to 60 degrees from the vertical shall have no harmful effect.
4	Water splashing against the enclosure from any direction shall have no harmful effect.
5	Water projected by a nozzle (6.3mm) against enclosure from any direction shall have no harmful effects.
6	Water projected in powerful jets (12.5mm nozzle) against the enclosure from any direction shall have no harmful effects.
7	Ingress of water in harmful quantity shall not be possible when the enclosure is immersed in water under defined conditions of pressure and time (up to 1 m of submersion).
8	The equipment is suitable for continuous immersion in water under conditions which shall be specified by the manufacturer. Normally, this will mean that the equipment is hermetically sealed. However, with certain types of equipment, it can mean that water can enter but only in such a manner that it produces no harmful effects.

Appendix D – About the Authors

Marshall Masters – Primary Author

Marshall Masters is a former science features news producer, freelance writer, television analyst, and publisher since 1999. A popular preparedness author, his numerous books address earth changes, space threats, and sustainable survival strategies and technologies. His current mission is to help survival community leaders create sustainable communication strategies using a wide range of affordable consumer and amateur two-way radios.

Seeing the need for affordable and effective off-the-shelf solutions, he began his research for this book two years prior to it's publication. He has had extensive volunteer training in all-hazards leadership programs which are offered by FEMA through the Amateur Radio Emergency Service (ARES) and Community Emergency Response Team (CERT).

Marshall holds an FCC amateur general class radio license. His FEMA training certifications include: Incident Command System (ICS) courses, 100, 200, 300, 400, 700 and 800, P-154, Rapid Visual Screening for Potential Seismic Hazards, ATC-20 Post earthquake Safety, and NIMS ICS All-Hazards Communications Unit Leader (0969).

Duane W. Brayton – Joint Author

As a young farm boy, Duane became fascinated with electronics and radios. When he was twelve years old, he began building shortwave radios from kits and would listen to them long into the night. After graduating high school in 1957, he joined the Air Force where he served honorably as a ballistic missile advanced electronics technician.

Following the Air Force, Duane began a career at 3M where he continued to build his electronics skills through industry conferences and classes. During his 21 years with 3M as a field service engineer, he was employed in various roles. Concurrent with his professional career at 3M, Duane also served over 20 years as a volunteer squadron commander and emergency communications officer for the Colorado Civil Air Patrol and as a fire lieutenant in the Cascade Colorado Volunteer Fire Department.

After he retired from 3M, he had another 15 year career as an electronics engineer for SCI System, United Circuits, and AGCO Printed Circuitry. He currently holds an amateur general class FCC license and enjoys building and testing HAM radios and antenna systems for survival applications.

Alphabetical Index